Intelligent Systems Reference Library

Volume 267

Series Editors

The aim of this series is to publish a Reference Library, including novel advances and developments in all aspects of Intelligent Systems in an easily accessible and well structured form. The series includes reference works, handbooks, compendia, textbooks, well-structured monographs, dictionaries, and encyclopedias. It contains well integrated knowledge and current information in the field of Intelligent Systems. The series covers the theory, applications, and design methods of Intelligent Systems. Virtually all disciplines such as engineering, computer science, avionics, business, e-commerce, environment, healthcare, physics and life science are included. The list of topics spans all the areas of modern intelligent systems such as: Ambient intelligence, Computational intelligence, Social intelligence, Computational neuroscience, Artificial life, Virtual society, Cognitive systems, DNA and immunity-based systems, e-Learning and teaching, Human-centred computing and Machine ethics, Intelligent control, Intelligent data analysis, Knowledge-based paradigms, Knowledge management, Intelligent agents, Intelligent decision making, Intelligent network security, Interactive entertainment, Learning paradigms, Recommender systems, Robotics and Mechatronics including human-machine teaming, Self-organizing and adaptive systems, Soft computing including Neural systems, Fuzzy systems, Evolutionary computing and the Fusion of these paradigms, Perception and Vision, Web intelligence and Multimedia.

Indexed by SCOPUS, DBLP, zbMATH, SCImago.

All books published in the series are submitted for consideration in Web of Science.

Dana M. Barry • Hideyuki Kanematsu
Editors

Applications of Metaverse and Virtual Reality to Creative Education and Industry

Springer

Editors
Dana M. Barry
Dept of Electrical & Computer Eng
Clarkson University
Potsdam, NY, USA

Hideyuki Kanematsu
Dept of Materials Science & Engineering
Natl Inst Technology, Suzuka College
Suzuka-shi, Mie, Japan

ISSN 1868-4394 ISSN 1868-4408 (electronic)
Intelligent Systems Reference Library
ISBN 978-981-96-3343-2 ISBN 978-981-96-3341-8 (eBook)
https://doi.org/10.1007/978-981-96-3341-8

This Springer imprint is published by the registered company Springer Nature Singapore Pte Ltd.
The registered company address is: 152 Beach Road, #21-01/04 Gateway East, Singapore 189721, Singapore

Preface

We, the editors Prof. Dana Barry and Prof. Hideyuki Kanematsu (who also serve as chapter authors), are proud to introduce our unique book *Applications of Metaverse and Virtual Reality to Creative Education and Industry*. This exciting book provides excellent examples of these technologies and their unlimited possibilities for creative education and industry. We have experience in industry as engineers and have had an interest in Creative Education for many years. Previously, our program to promote creative education globally was awarded a National Chem-luminary Award from the American Chemical Society (ACS) in 2004. We have successfully carried out many creative problem-based learning activities in Second Life, SL (a three-dimensional world where avatars perform certain tasks on behalf of us). Some of these activities are described in this book. On the other hand, metaverse provides a fully immersive and interconnected virtual world where individuals can work, live, etc. and interact in a digital environment that closely resembles the real world.

Creative education is important to motivate and engage students into active and enjoyable learning. It is also important for turning students on to STEM (science, technology, engineering, and mathematics) courses and careers to supply our future scientists, engineers, etc., who identify and help solve the challenging problems of the world. These professionals are essential to various industries that provide us with food, medicine, and more. Therefore, creative, and enthusiastic teachers are needed. Barry successfully presented Creative Teaching Workshops to teachers in SL. She provided them with unique teaching approaches (e.g., the multisensory method, which incorporates the senses into lessons to captivate the learning styles of all students). Then she asked them to prepare creative lessons to use with their students and to share them with her SL workshop attendees.

We have taught and continue to teach exciting student lessons using virtual reality (VR) and (VR) headsets. Our research results show that in general, students enjoy, learn, and prefer a 3-D teaching method over the use of a traditional textbook. VR headsets use software that allows individuals to interact with real to imaginary computer simulated environments. The 3-D experiences include both visual and auditory components. One example is a scary and thrilling visit to Jurassic Park, where you are surrounded by many roaring and fighting dinosaurs, etc. This VR experience could be used for engaging and exciting tours of Jurassic Park, for a lesson about dinosaurs, and for a study of interesting time periods (Jurassic Period, Mesozoic Era, Triassic Period, etc.). In addition, VR headsets combined with JINS MEME sensor glasses are used to obtain physiological data (head movement, eye blinking counts, etc.) from individuals using the VR headsets. This provides a measurement for students' reactions such as excitement or surprise.

Our book presents the exciting technology of virtual reality and its many benefits for creative education and industry. Several benefits include the following. It allows students to work from anywhere in the world, at any time, and at their own pace. It is advantageous to design and make houses, etc. in SL because expensive building materials are not necessary and potential safety risks are avoided. Such design techniques may be used in various industries for the construction of new buildings and new products. Another related topic is the importance of virtual reality to architectural design because it allows architects to create immersive 3-D environments to share with their clients and customers. Also, we mention the application of virtual reality to medical education. One example is the use of simulations to provide learners with a safe and immersive environment to practice their skills and knowledge. Another book section describes how the rapid advancement of technology (like artificial intelligence, machine learning, the internet of things, augmented reality, robotics, etc.) is reshaping industries, societal norms, manufacturing, education, and more. Another interesting chapter describes a method used to build virtual worlds for the video game industry. Virtual worlds provide the setting and background for video games. Most of us have played video games and some of us are designing and marketing them. The final chapter is a projection for the future, where the real and virtual worlds fuse together. It discusses the integration of Digital Twin technology with augmented reality which is transforming industries by bridging the gap between the physical and digital worlds.

This book serves as an excellent textbook, as well as a reference book and practical guide. It may be of special interest to professors, scientists, and engineers of diverse fields and their graduate students. Also, this book may attract the public and individuals who want information about metaverse, avatars, virtual reality, VR headsets, and other related topics.

We invite you to explore the fascinating world of avatars, metaverse, and virtual reality through this book. It promises to be a rewarding journey revealing the limitless potential of these technologies for creative education and industry.

Very Best Wishes for the Present and the Future,

Potsdam, NY, USA — Dana M. Barry
Suzuka-shi, Mie, Japan — Hideyuki Kanematsu

Acknowledgments

We thank the Japanese government and the headquarters of KOSEN (National Institute of Technology, Japan) for supporting our research activities to develop anti-infective materials. Particularly, we thank Dr. Isao Taniguchi, the President of the National Institute of Technology (KOSEN), Japan, from the bottom of our hearts, for his encouragement and support. Under the GEAR 5.0 Project of KOSEN, we thank the Senior Executive Directors of KOSEN, Professor Eiji Takada (Senior Executive Professor) and Professor Yasuo Utsumi (Specially Appointed Professor for the GEAR 5.0 Project) who have also encouraged us to pursue our international projects in various ways. We greatly appreciate them. We are very thankful to Clarkson University, the Dean of Clarkson's Wallace H. Coulter School of Engineering Professor William Jemison, Clarkson's Department of Electrical & Computer Engineering, and the Chair of this Department, Professor Paul McGrath. We also thank the State University of New York at Canton, the National Institute of Technology, Suzuka College in Japan, and its President Professor Shinji Fujimoto. Thanks are extended to Executive Vice President of Osaka University Professor Toshihiro Tanaka, Professor Eiji Arai of Osaka University, Professor Takayoshi Nakano, Osaka University in Japan, the American Chemical Society, Professor Dr. Roger Haw Boon Hong, and Ansted University, for their greatly appreciated support.

We would like to thank several academic societies related to Information and Communication Technology (ICT) and engineering education, including KES International, the American Society for Engineering Education (ASEE), and e-Help (e-Learning Higher Education Linkage Project) in Japan. We are also grateful to the leading materials science and surface finishing societies—including The American Society for Metals (ASM) International, TMS (The Minerals, Metals & Materials Society), ACerS (The American Ceramic Society), AIST (Association for Iron & Steel Technology), and NASF (National Association for Surface Finishing) in the United States; the Institute of Materials Finishing (IMF) in the United Kingdom; and Japan's Institute of Metals and Materials (JIM), the Materials Research Society of Japan (MRS-J), and the Surface Finishing Society of Japan (SFSJ)—which have encouraged us to integrate digital scientific components into materials science.

Some information contained in this book was obtained from work carried out for national funding projects in Japan. (Particularly, we appreciate JSPS Kakenhi 19K12246, 21H00914, 22K12284, etc. in Japan and enthusiastic colleagues all over the world related to these funding projects.)

Special thanks are extended to the chapter authors for their excellent contributions to this book. Those from Japan include Professor Nobuyuki Ogawa of the National Institute of Technology at Gifu College, Professor Hideyuki Kanematsu of the National Institute of Technology at Suzuka College, Professor Yutaka Tada of the National Institute of Technology at Anan College, Professor Takashi Matsumoto of the National Institute of Technology at Anan College, Professor Takahisa Kamikura at Suzuka University of Medical Science, and Professor Inaba of the Department of Emergency Medicine at Kanazawa Medical University. The authors in the United States are Chair Professor Paul McGrath of the Department of Electrical and Computer Engineering at Clarkson University, Research Professor Dana Barry of the Department of Electrical and Computer Engineering at Clarkson University, and Morgan Hastings, Lecturer and Program Director for Graphic and Multimedia Design at the State University of New York at Canton.

We also express our sincerest thanks and appreciation to Professor Dr. Lakhmi Jain, Series Editor for the Intelligent Systems Reference Library, Editor Dr. Mei Hann Lee, and the entire Springer team.

In addition, we would like to thank our families for their continued interest and support: Dr. Barry's parents (Daniel and Celia Malloy), her husband (James), and their four sons (James and his wife Linda and their children Alex, Jonathan, Christopher, and Catherine), (Brian and his wife Erica), (Daniel), and (Eric and his wife Karem and their son Bodie). Also we thank Dr. Kanematsu's parents (Shoji and Michiko Kanematsu), his wife (Reiko), and their children (Hitomi and Hiroyuki).

Dana M. Barry
Hideyuki Kanematsu

Contents

About the Editors

Dana M. Barry, Ph.D., is a Research Professor in the Department of Electrical & Computer Engineering at Clarkson University in the United States and an Instructional Support Assistant at the State University of New York at Canton. She is a Professor and Scientific Board President for Ansted University and has served as a visiting professor overseas, numerous times. So far, she has been invited as either or both a keynote speaker and an invited guest speaker in the following countries: the United States, Malaysia, Japan, China, Slovenia, Serbia, France, Greece, Spain, England, Hungary, Poland, and Italy.Professor Barry earned B.A. and M.S. degrees in Science Education from the State University of New York at Potsdam (SUNY Potsdam), and an M.S. degree in Chemistry from Clarkson University. In addition, she has three doctoral degrees including a Ph.D. in Engineering from Osaka University, Japan. Dr. Barry is a member and chemistry ambassador for the American Chemical Society and has served (for over 20 years) as a coordinator for both National Chemistry Week and National Earth Week for the Northern New York Section of the American Chemical Society, along with serving as a section officer for over 25 years. She belongs to the Planetary Society and has pursued some studies in space exploration. She has her name on the Mars Rovers (Spirit, Opportunity, and Curiosity), the MAVEN spacecraft that flew to Mars, the NASA spacecraft that landed on the asteroid Bennu in 2020, and on the Japanese spacecraft Hayabusa -2, the first one in the world to land on an asteroid. She has over 400 academic publications including six Springer Nature textbooks (with most of them being listed as the best reference books in the world). Dr. Barry served as Coauthor and Coeditor for these Springer books. One of her many honors includes serving as host and co-producer of her own television show (Sensational Science) that aired for several years in Northern New York State. Other honors include winning 21 consecutive APEX Awards for Publication/Communication Excellence from Communications Concepts in Springfield, VA; a Marquis Who's Who Lifetime Achievement Award for her outstanding professional contributions (2017); an Outstanding Volunteer Award from the American Chemical Society (2017); and a National Chem-Luminary Award from the American Chemical Society for promoting chemistry at the international level. In 2019, Dr. Barry received an Outstanding

Achievement Award from the Materials Research Society of Japan (MRS-J) for outstanding work involving biofilms and contamination control on materials' surfaces. Also, in 2021, she was nominated by the Academic Union of Oxford, UK, to receive the special title "The Name in Science" and a personal medal. This award is for the most respected personalities in the fields of science, technology, and innovations. In addition, Dr. Barry received the Leon LeBeau award (November 13, 2024) for her outstanding and long-lasting contributions to the members of SOAR (a large organization in Northern New York State). Her expertise is in science and creative education, STEM (science, technology, engineering, and mathematics) education, chemistry, engineering, and for her collaborative work with Dr. Kanematsu for biofilms, antiviral materials, etc., problem-based learning activities in Second Life, and virtual reality.

Hideyuki Kanematsu, Ph.D., Professor, is a distinguished researcher in materials science and engineering, specializing in surface science and biofilm countermeasures. He is a Professor at the National Institute of Technology (NIT), Suzuka College, in Japan, and serves as a visiting professor at Nagoya University. He holds Fellowships with ASM International (FASM, USA) and the Institute of Materials Finishing (FIMF, UK). Dr. Kanematsu earned his B.Eng., M.Eng., and Ph.D. in Materials Science and Engineering from Nagoya University, Japan.Throughout his career, Dr. Kanematsu has contributed significantly to both academia and industry, serving as Dean of the Department of Materials Engineering and as Deputy President at NIT, Suzuka College. He currently directs the Materials Science and Engineering Center under Japan's GEAR 5.0 project, which advances research on antibacterial, antiviral, and antibiofilm materials. His research examines interfacial phenomena between materials and microorganisms, aiming to develop innovative surface treatments with applications in healthcare and environmental protection. Dr. Kanematsu has authored over 630 publications, including co-authoring four notable Springer textbooks. His work has earned numerous accolades, including Japan's Minister of Education Award in 2021. Recognized internationally, he is advancing studies on 3D-printed surface modifications to enhance material performance against biofilms, positioning him as a leader in his field. In addition, he belongs to the Institute of Materials and Surface Finishing (IMF), the American Society for Materials Science (ASM International), the American Chemical Society (ACS), the American Ceramic Society (ACerS), Japan Institute of Metals (JIM), the Iron and Steel Institute of Japan (ISIJ), the Surface Finishing Society of Japan (SFSJ), the Electrochemical Society of Japan (ECSJ), MRS-J, Japan Society of Thermal Treatment Technology (JSHT), Japan Thermal Spray Society (JTSS), and other academic societies in Japan and abroad.

Chapter 1
Creative Education and Its Importance to Industry

Dana M. Barry, Hideyuki Kanematsu, and Paul McGrath

Abstract This chapter starts by defining creativity as the ability to produce original ideas and items. It also includes the combining of existing work and objects in different ways for new purposes. Creativity involves higher levels of thinking like the synthesis level (where one creates new ideas) and is important to all areas of education. Every field of study has problems to solve and relies on creative ideas for possible solutions. Combining creativity with education is creative education. It encourages students to be open-minded and creative. Engineering education is very important for providing trained and innovative engineers to work in industry. Engineers along with other STEM (science, technology, engineering, and mathematics) graduates are needed to creatively design items, processes, and services to satisfy human needs. Industrial engineers use a design process (a problem-solving model) to solve problems for meeting the demands of society and to compete globally in areas like space exploration. This chapter provides details about creativity and creative education. It also describes various applications of creative education to industry.

Keywords Creativity · Creative thinking · Creative education · STEM (science, technology, engineering, & mathematics) · Industry

D. M. Barry (✉) · P. McGrath
Clarkson University, Potsdam, NY, USA
e-mail: dbarry@clarkson.edu; pmcgrath@clarkson.edu

H. Kanematsu
National Institute of Technology, Suzuka College, Suzuka, Mie, Japan
e-mail: hideyuki.kanematsu@bioenglab.org

D. M. Barry, H. Kanematsu (eds.), *Applications of Metaverse and Virtual Reality to Creative Education and Industry*, Intelligent Systems Reference Library 267, https://doi.org/10.1007/978-981-96-3341-8_1

1.1 Creativity

Creativity has been interpreted in many ways throughout history. In this book, it is described as the capacity to generate novel ideas, concepts, or products, including tangible items such as toys and medicine. It also encompasses reimagining and combining existing works, objects, and ideas in innovative ways to serve new purposes. Combining creativity with education is creative education. It encourages students to be open-minded and creative thinkers. Creativity includes the creative person, creative process, and the creative product. Barriers to creativity also exist.

Creativity has had a variety of meanings over the years. In the context of this book, creativity is defined as the ability to generate unique ideas [1], create new items, and develop products such as toys and medicine. It also involves blending existing concepts, objects, and works in innovative ways to achieve new goals. A simple example is the wheelchair, which is the result of combining wheels with a chair. Creativity engages advanced cognitive processes, including synthesis, as outlined in Bloom's Taxonomy [2]. Bloom's Taxonomy is a pyramid with six cognitive categories (listed in order of the lowest level to the highest level: knowledge, comprehension, application, analysis, synthesis, and evaluation). At the knowledge level, students memorize and recall information. They grasp ideas at the comprehension stage and implement rules at the application stage. Numerous engineering challenges fall under the analysis stage, where students typically evaluate a single technical solution. Individuals create ideas, etc. at the synthesis level and make value judgments, etc. at the evaluation level.

Creativity starts with a creative person using a creative process to make a creative (new) product [3]. A creative person is usually full of ideas and energetic. This person exhibits a strong drive for personal growth, spontaneity, divergent thinking, openness to novel experiences, persistence, and a diligent work ethic [4].

Researchers have conducted studies to better understand the behavior of creative individuals. Chavez-Eakle, Lara, and Cruz identified that creative individuals tend to explore new situations with enthusiasm, display optimism, tolerate ambiguity, and pursue their objectives with determination [5]. Additionally, the Torrance Tests of Creative Thinking (TTCT) have been widely used to measure creative potential. These tests evaluate creativity through indices and scores in categories such as fluency, originality, elaboration, and flexibility [6]. Fluency measures the total number of meaningful ideas generated, while originality assesses the uniqueness of responses. Elaboration evaluates the level of detail provided, and flexibility considers the variety of different categories within the responses.

For decades, scholars have analyzed the cognitive strategies of renowned figures such as Thomas Edison, the inventor of the light bulb and one of the most prolific inventors of the late nineteenth century. Famous scientists and inventors such as Edison and Einstein are often seen as highly productive thinkers. When they face a problem, they will first think about how to approach the problem. Their ability to identify a solution comes from their skill for connecting unrelated concepts.

NOTE: Individuals can promote creativity by daydreaming with a purpose, jotting down notes while brainstorming ideas, practicing divergent thinking, and by connecting unrelated words, items, etc.

Happiness can promote creativity. For example, when people are in positive moods due to joyfulness, hope, etc., they are motivated to explore and accept new ideas and new information. On the other hand, it should be mentioned that a few studies have been carried out to determine the positive effect of creativity on subjective wellbeing (a focus on both positive and negative aspects of life as well as satisfaction with life) [7]. One such study was carried out with students and adults in Malaysia. The results imply that creativity is positively associated with subjective well-being, which is a broader and more comprehensive measure of wellbeing. This research is presented in a publication by Cher Tan, Chuah, Lee, and Chee Tan [7–24].

The creative process includes the thinking and the acts that take place to produce an original item [25]. This process involves identifying a problem, searching for information and new methods to get potential solution ideas, and solving the problem. The Cambridge Handbook of Creativity summarizes some theories of creativity [26]. A few are mentioned. Psychometric theory measures creativity in terms of assessing the reliability and validity of the creative product [3]. Economic Theories are concerned with creative ideas and behavior that are influenced by the marketplace and the economy. Cognitive Theories relate to creative people who use remote associations, divergent and convergent thinking, conceptual combinations, and metacognitive processes (which relate to the awareness of one's own knowledge). While many models exist for the creative process, most of them include steps with the common themes of synthesis, analysis, and evaluation (which are the higher levels of thinking). Scientists are still struggling to describe the creative process. Dr. Adam Green, a cognitive neuroscientist at Georgetown University feels that the process involves coupling different regions of the brain.

Creativity often involves coordination between the cognitive control network (which includes executive functions like planning and problem-solving) and the default mode network, which is active during mind-wandering or daydreaming. Green found evidence that an area called the frontopolar cortex, in the brain's frontal lobes, is related to creative thinking [27–54]. He and his colleagues used transcranial direct current stimulation (t-DCS) to stimulate the frontopolar cortex of participants trying to come up with novel analogies. Stimulating this area made the participants act more creatively. More studies are needed to further develop this research topic.

The creative product is one that never existed before. It could be a new book, song, toy, or invention. Figure 1.1 shows a child's toy horse on wheels from Ancient Greece [55]. Figure 1.2 displays a creative flower [56]. A creative game to play and an innovative recipe for candy are also creative products. Other examples include Einstein's Theory of Relativity and Shakespeare's plays. Products like publications and musical compositions can be counted and are often available for viewing. Some creative items like Michelangelo's paintings remain popular for a long time, while others have never been socially noticed.

Fig. 1.1 This figure shows a child's toy horse on wheels from Ancient Greece

Fig. 1.2 A creative flower

There are always obstacles to creativity. However, in order to solve various problems, it is necessary to recognize the obstacles and raise the level of thinking necessary to solve various problems. These obstacles may include gaps in knowledge, limitations, emotional resistance, cultural constraints, and poor expression. To create innovative solutions, students will need to acquire sufficient knowledge, while emotional resistance, such as a fear of failure, will need to be overcome in order to increase their willingness to think without hesitation using new approaches.

1.2 Creative Education

Educating for creativity is the essence of creative education. It is important here that students are encouraged to learn actively and with enjoyment. It is also needed for turning students onto STEM (science, technology, engineering, and mathematics) courses and careers to provide our present and future scientists, engineers, etc. who identify and help solve the challenging problems of the world. These professionals are essential to various industries that provide us with food, medicine, and more. Therefore, teachers need to be creative and enthusiastic. They must be open-minded, positive about the topics they teach, and value originality. These expressed traits encourage students to share their creative ideas.

The editors Barry and Kanematsu have experience in industry as engineers and have had an interest in Creative Education for many years. Previously, their program to promote creative education globally was awarded a National Chem-Luminary Award from the American Chemical Society (ACS) in 2004. It should be mentioned that Prof. Dr. Roger Haw of Malaysia helped promote the program in China and Malaysia. Figure 1.3 shows Dr. Barry receiving the award on behalf of the Northern New York Section of the American Chemical Society at the 2004 ACS Convention in Philadelphia [57]. The winning program includes creative teaching methods, books, and lessons developed by Barry. These materials have been used by many

Fig. 1.3 Dr. Barry receives the Chem Luminary Award at the ACS Convention in Philadelphia

instructors and hundreds of students worldwide. Descriptions of the creative teaching methods are provided.

The Multisensory Teaching Approach (also referred to as the Chemical Sensation Project) encourages educators to integrate sensory experiences (such as seeing, hearing, smelling, touching, and tasting) into science lessons and engineering design activities [58]. This approach was created to address the diverse learning styles of students. Some students learn best through visual input, others through auditory methods, and some by engaging in hands-on activities. Using this method, students are presented with a problem to solve and are motivated to work independently. For instance, they might be tasked with designing devices to create high-quality popcorn. In this scenario, the devices are assessed based on the popcorn they produce. The sense of sight is used to judge the visual appeal of the popcorn, while the senses of smell and taste evaluate its aroma and flavor.

The Science Fair Project Teaching Approach to creative education allows students to select a theme of interest from among the science fair and design engineering challenges and solve the problem [59]. During the project, students think broadly and use a variety of communication tools, including charts, graphs, and oral and written presentations. Students are expected to work independently as they gather, analyze, and interpret data to draw conclusions for their exciting projects. For example, a university student in Japan selected a project focused on studying the growth of bubbles, as illustrated in Fig. 1.4 [60].

The Reading and Solving a Mystery Teaching Approach encourages students to take on the role of a detective, solving a mystery (such as a crime or problem) by gathering and analyzing evidence (data). This method incorporates published books written in English and Japanese by Barry and Kanematsu, which feature mystery stories and related activities in science and engineering [61]. In this approach,

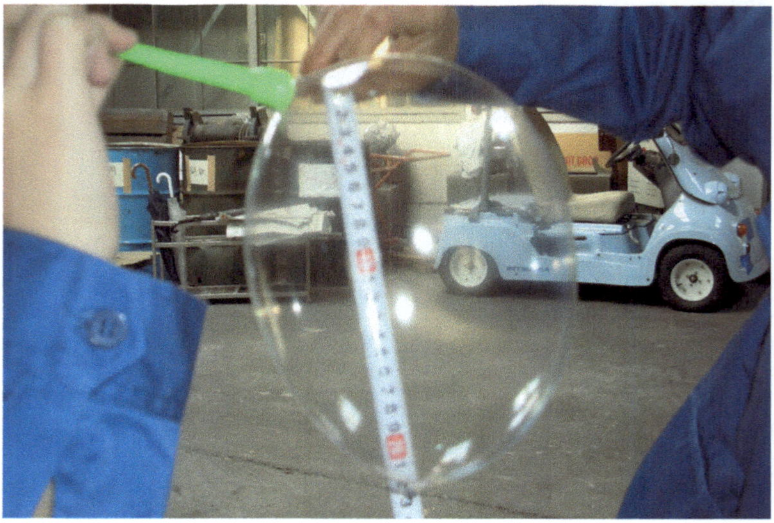

Fig. 1.4 A college student in Japan closely examines the appearance and size of a bubble

students read a mystery story and treat it as a research project. They identify the central problem or crime, gather evidence (data), and creatively develop and execute an investigation to resolve the issue. This method is adaptable and can be applied to various mystery stories.

The Space-Related/Space Exploration Teaching Approach fosters communication skills, creativity, and planning abilities in science and engineering students. It encourages them to visualize unknown objects and scenarios, particularly when engaging in science and engineering design projects set beyond the confines of planet Earth.

One project might be to predict the properties of an unknown planet appearing between Earth and Mars. Knowing some information about Earth and Mars, an individual could make an educated guess about the possible properties of the unknown planet. Let's take the global environment as an example. It is well known that the Earth has oceans and an atmosphere, and that approximately 78% of the atmosphere is composed of nitrogen and approximately 21% is composed of oxygen. On the other hand, what about the atmosphere of Mars? The atmosphere of Mars is composed of carbon dioxide. Mars is a smaller planet than Earth, covered in rocks, and has ice at the poles and volcanoes. In this educational approach, students work to solve problems under the guidance of a lecturer acting as a facilitator. Teams made up of students from diverse backgrounds brainstorm and discuss the target problem to find a solution through cooperation. They create multiple sketches, diagrams, and other visuals to communicate their ideas. This approach can also be effectively applied to projects unrelated to space exploration.

The Innovative E-learning and Team Approach for Problem Solving in Virtual Reality incorporates software tools and digital media. It offers flexibility, allowing portions or the entirety of a project to be conducted through e-learning platforms. This framework enables participants to engage in simulations such as Mars Explorer Missions, and problem-based learning (PBL) activities within the Metaverse (a three-dimensional environment like Second Life) where avatars perform certain tasks on behalf of the users. Figure 1.5 shows Numazu college students in Japan participating in a Mars Explorer simulation mission [60]. Using real data on Mars, students choose a landing site and launch a spaceship equipped with a rover. Then, the students explore Mars in a 3D virtual space. In this educational method, each participant uses simulation software to move the rover and work in groups to solve problems. The groups hold brainstorming sessions to generate possible solutions and select the best one. An example of PBL involves students designing and constructing a futuristic house using primitives (three-dimensional virtual objects such as cubes) within the virtual environment of Second Life. Barry and Kanematsu have effectively implemented numerous innovative problem-based learning projects with their students in Second Life. Some of the activities are described in this book.

TWO OTHER CREATIVE TEACHING STRATEGIES ARE PROVIDED [62]

The Brainstorming Strategy is meant to excite the mind to think in all directions about all possibilities for a problem or topic. This is a great classroom activity to get

Fig. 1.5 Numazu college students participate in a Mars Explorer simulation mission

the flow of students' ideas and mental creations. It also provides opportunities for the students to be active learners, to respect different opinions, and to take advantage of other people's ideas.

The Discussions and Debates Strategy requires the instructor to ask thought-provoking questions to inspire students to think for themselves. They also need to express their unique opinions in a discussion. A debate can be held like a competition with teams and run by the teachers. It is a good way to motivate students to learn more about the topics being discussed because they need to research them to defend their point of view and win the debate. This type of activity enhances critical thinking skills and generates new ideas and solutions.

ANOTHER CREATIVE TEACHING METHOD IS CONSIDERED

Combining Artistic and Engineering Disciplines may benefit Students. Professors Innella and Rodgers are design practitioners, researchers, and educators, who are promoting the potential benefits of combining engineering and art research [63]. They encourage an education that merges arts and humanities disciplines with scientific and technical subjects. During the Renaissance (1300–1700), artistic and scientific research seemed to complement each other. The intellectual man of the Renaissance was considered a "polymath." A polymath is a person whose talents include art and engineering, design, and mathematics, as well as philosophy and science. Many considered Leonardo da Vinci to be a polymath who was active as a painter, draftsman, engineer, scientist, theorist, sculptor, and an architect. One of his famous paintings is the Mona Lisa.

With the passing of many years, most schools display a division between artistic and engineering disciplines. However, examples exist of organizations bringing the art and engineering worlds together through interdisciplinary teams and projects, etc. [60, 62–73]. More recently, the importance of the arts has gained attention along with the creation of the acronym STEAM (science, technology. engineering, art, and mathematics).

People supporting the addition of the arts to a group called STEM, see it as an opportunity to enhance students' soft skills like aesthetics, real world applications, and communication. This is especially important because our world is increasingly ruled by technology like algorithms, machine learning, robots, and artificial intelligence (AI), etc. Digital techniques have brought creative ideas to fruition and modern creativity often relies on them. Artificial intelligence often excels where humans struggle. Therefore, a collaboration of humans with machines can provide positive results where diversity and innovation flourish [74]. However, individuals must continue to develop their technical, communication, and other skills. Robots can work independently. Industrial robots are programmable. They are also autonomous machines capable of carrying out a complex set of operations (tasks that are often hazardous to humans). On the other hand, collaborative robots are called co-bots. They work alongside human workers in the same workspace. They are flexible. Therefore, they can be used for multiple tasks, so they are adaptable to changing work requirements. This type of collaboration often uses sensors and other forms of safety. These measures are used to make sure the co-bots operate securely and effectively without the need for fences, cages, etc.

1.3 Applications of Creative Education to Industry

Creative education encourages students to be open-minded and creative. Creative engineering education is very important for providing trained and innovative engineers to work in industry. Industrial engineers use a design process (a problem-solving model) to solve problems for meeting the demands of society and to compete globally in areas like space exploration. A typical engineering design process includes the following steps. (1) Identify the problem. (2) Collect data to solve the problem. (3) Identify design requirements. (4) Identify design limits. (5) Generate possible solutions to the problem. (6) Evaluate the alternatives. (7) Select the best approach. (8) Communicate the selected design. (9) Implement the design. (10) Test the product. (11) Modify the design if needed. Engineers, including other STEM graduates where STEM stands for science, technology, engineering, and mathematics, are needed to creatively design items, processes, and services to satisfy human needs. These professionals (many of whom work in industry) identify and help solve the challenging problems of the world.

Creative education is very important to industry. Chunfang Zhou mentions that creativity is one key capability that engineering students should master. He believes that Creative Engineering Education is needed to foster creative students to face the challenges of complex engineering work [75]. Several other authors provide creative tips and information to include in Creative Education programs for engineering students to benefit industries and society. C. Selinger (a creative engineer) suggests that the use of synectics is better than the typical brainstorming session [76]. Synectics is a brainstorming discussion including individuals involved with the project and a representative of the group (client) asking them to address the problem. Each person writes down a dozen or more "I wish" statements and shares them with the group so others can build on them and come up with more ideas. No one is allowed to criticize these ideas because it would discourage creativity. The ideas are organized into groups and reduced to a manageable number of options. The client takes notes and has the authority to act on any feasible ideas.

Good Creative Education programs benefit industry and its scientists, engineers, etc. by helping them become more creative with new ideas for improving the products that they design and build (make) [77–81].These programs encourage individuals to look at things from different perspectives and to regularly talk to people from different industries, cultures, and professions, and to their customers to get new ideas. Oliver Broadbent (Director of Constructivist Ltd.) believes that creative thinking is just as important to Civil Engineering education as analytical and technical thinking [8]. He feels that much of the training previously focused on convergent areas of thinking like analysis, construction, material properties, and how to reach performance and safety requirements. He mentions that it is important to know what to build before you build something. This involves a whole series of thought processes, creative thinking with divergent activities.

Creative industries should be mentioned. They include the production of music, the performing and visual arts, films, computer games, etc. These industries usually have creative founders (those who have natural creative ability and/or those who have participated in creative education programs). The creative founders mentioned in the study by Loots and Bennekom do not strive for profit maximization [82–97]. They proactively seek peer, consumer, and expert recognition, which often results in growth and success. If possible, they try to create it on their own. Creative industries use novel ideas within social networks for production and consumption. Another study was carried out to find out how social factors affect an individual's creativity [88–94]. Data was obtained by analyzing over 200 valid questionnaires completed by creative entrepreneurs in creative enterprises. At the personal level, creative entrepreneurs share two perceptions in society (self-perception and social perception). When self-perception is emphasized, creative entrepreneurs' creativity focuses on personal goals and needs. However, when social perception is emphasized, the entrepreneurs' creativity focuses on collective norms and values. The results also show that individual creativity has a positive impact on leadership.

As one can imagine, Creative Education programs continue to grow and modify their components to prepare students to be creative thinkers for meeting the needs of society. Our global community is constantly changing and relying more on advanced technologies. Therefore, these changes and use of advanced technologies need to be incorporated into Creative Education programs that prepare the present and future engineers, scientists, etc. working in industry and elsewhere to identify and solve the challenging problems of the world.

1.4 Conclusions

This chapter starts by defining creativity, the ability to produce original ideas and items. It also includes the combining of existing work and objects in different ways for new purposes. Creativity involves higher levels of thinking like the synthesis level (where one creates new ideas) and is important to all areas of education. Every field of study has problems to solve and relies on creative ideas for possible solutions. Combining creativity with education is creative education. It encourages students to be open-minded and creative. Engineering education is very important for providing trained and innovative engineers to work in industry. Engineers along with other STEM (science, technology, engineering, and mathematics) graduates are needed in industry and elsewhere to creatively design items, processes, and services to satisfy human needs. Industrial engineers use a design process (a problem-solving model) to solve problems for meeting the demands of society and to compete globally in areas like space exploration. This chapter provides details about creativity and creative education, which continues to grow and modify its components. Our global community is constantly changing and relying more on advanced technologies. Therefore, these changes and uses of advanced technologies need to be incorporated into Creative Education programs that prepare the present and future engineers, scientists, etc. to identify and solve the challenging problems of the world. The chapter also describes some applications of creative education to industry.

References

1. Smith, S.: Exploring Collaborative Design in a PK-12 Creative Writing Challenge. Doctoral dissertation (2012)
2. Bloom, B.S., Krathwoh, D.R.: Taxonomy of Educational Objectives: The Classification of Educational Goals by a Committee of College and University Examiners. Handbook 1: Cognitive Domain. Longmans, New York (1956)
3. Kanematsu, H., Barry, M., Kanematsu, D.H., Barry, D.M.: Theory of creativity. In: STEM and ICT Education in Intelligent Environments, pp. 9–13 (2016)
4. Anderson, H.: Creativity in perspective. In: Creativity and Its Cultivation, pp. 236–242. Harper Brothers, New York (1959)
5. Chavez-Eakle, R.A., Lara, Cruz, C.: Personality: a possible bridge between creativity and psychopathology. Creat. Res. J. **18**(1), 27–38 (2006)
6. Torrance, E.P.: Torrance Tests of Creative Thinking. Scholastic Testing Service, Bensenville (1990)
7. Tan, C.-Y., Chuah, C.-Q., Lee, S.-T., Tan, C.-S.: Being creative makes you happier: the positive effect of creativity on subjective well-being. Int. J. Environ. Res. Public Health. **18**(4), 7244 (2021)
8. Fiorelli, J.A., Russ, S.W.: Pretend play, coping, and subjective well-being in children: a follow-up study. Am. J. Play. **5**, 81–103 (2012). https://doi.org/10.1037/e700772011-001
9. Tan, C.S., Qu, L.: Stability of the positive mood effect on creativity when task switching, practice effect, and test item differences are taken into consideration. J. Creat. Behav. **49**, 94–110 (2015). https://doi.org/10.1002/jocb.56

10. Fujiwara, D., Lawton, R.N.: Happier and more satisfied? Creative occupations and subjective well-being in the United Kingdom. Psychosociol. Issues Hum. Resour. Manag. **4**, 50–74 (2016). https://doi.org/10.22381/PIHRM4220163
11. Tamanoir, M.R., Motaghedifard, M.: Subjective well-being and its sub-scales among students: the study of role of creativity and self-efficacy. Think. Skills Creat. **12**, 37–42 (2014). https://doi.org/10.1016/j.tsc.2013.12.003
12. Zhang, Y., Li, J., Song, Y., Gong, Z.: Radical and incremental creativity: associations with work performance and well-being. Eur. J. Innov. Manag. (2020). https://doi.org/10.1108/EJIM-12-2019-0351
13. Shao, Y., Zhang, C., Zhou, J., Gu, T., Yuan, Y.: How does culture shape creativity? A mini review. Front. Psychol. **10**, 1219 (2019). https://doi.org/10.3389/fpsyg.2019.01219
14. Tan, C.S., Tan, S.A., Cheng, S.M., Hashim, I.H.M., Ong, A.W.H.: Development and preliminary validation of the 20-item Kaufman Domains of Creativity Scale for use with Malaysian populations. Curr. Psychol. **40**, 1946–1957 (2021). https://doi.org/10.1007/s12144-019-0124-8
15. Hennessey, B.A., Amabile, T.M.: Creativity. Annu. Rev. Psychol. **61**, 569–598 (2010). https://doi.org/10.1146/annurev.psych.093008.100416
16. Diedrich, J., Benedek, M., Jauk, E., Neubauer, A.C.: Are creative ideas novel and useful? Psychol. Aesthet. Creat. Arts. **9**, 35–40 (2015). https://doi.org/10.1037/a0038688
17. Tan, C.S., Ong, A.W.H.: Psychometric qualities and measurement invariance of the modified self-rated creativity scale. J. Creat. Behav. **53**, 593–599 (2019). https://doi.org/10.1002/jocb.222
18. Diedrich, J., Jauk, E., Silvia, P.J., Gredlein, J.M., Neubauer, A.C., Benedek, M.: Assessment of real-life creativity: the Inventory of Creative Activities and Achievements (ICAA). Psychol. Aesthet. Creat. Arts. (2017). https://doi.org/10.1037/aca0000137
19. Tan, C.S., Lau, X.S., Lee, L.K.: The mediating role of creative process engagement in the relationship between shyness and self-rated creativity. J. Creat. Behav. **53**, 222–231 (2019). https://doi.org/10.1002/jocb.173
20. Tan, C.S., Lau, X.S., Kung, Y.T., Kailsan, R.: Openness to experience enhances creativity: the mediating role of intrinsic motivation and the creative process engagement. J. Creat. Behav. **53**, 109–119 (2019). https://doi.org/10.1002/jocb.170
21. Kaufman, J.C.: Counting the muses: development of the Kaufman Domains of Creativity Scale (K-DOCS). Psychol. Aesthet. Creat. Arts. **6**, 298–308 (2012). https://doi.org/10.1037/a0029751
22. George, J.M., Zhou, J.: When openness to experience and conscientiousness are related to creative behavior: an interactional approach. J. Appl. Psychol. **86**, 513–524 (2001). https://doi.org/10.1037/0021-9010.86.3.513
23. Zhou, J., George, J.M.: (2001) When job dissatisfaction leads to creativity: encouraging the expression of voice. Acad. Manag. J. **44**, 682–696 (2001). https://doi.org/10.5465/3069410
24. Kaufman, J.C., Beghetto, R.A.: Beyond big and little: the four C model of creativity. Rev. Gen. Psychol. **13**, 1–12 (2009). https://doi.org/10.1037/a0013688
25. Barry, D., Kanematsu, H.: Workshops in creative education for students and teachers in the United States and Japan. In: 2007 37th Annual Frontiers in Education Conference-Global Engineering: Knowledge Without Borders, Opportunities Without Passports, pp. T2B-16. IEEE (2007)
26. Kozbelt, A., Beghetto, R.A., Runco, M.A.: Theories of creativity. In: Kaufman, J.C., Sternberg, R.J. (eds.) The Cambridge Handbook of Creativity, pp. 20–47. The Cambridge University Press, New York (2010)
27. Green, A.E., Spiegel, K.A., Giangrande, E.J., Weinberger, A.B., Gallagher, N.M., Turkeltaub, P.E.: Thinking Cap plus thinking zap: tDCS of frontopolar cortex improves creative analogical reasoning and facilitates conscious augmentation of state creativity in verb generation. Cereb. Cortex. **27**(4), 2628–2639 (2017)
28. Aron, A.R., Robbins, T.W., Poldrack, R.A.: Inhibition and the right inferior frontal cortex: one decade on. Trends Cogn. Sci. **18**(4), 177–185 (2014)

29. Badre, D., D'Esposito, M.: Functional magnetic resonance imaging evidence for a hierarchical organization of the prefrontal cortex. J. Cogn. Neurosci. **19**(12), 2082–2099 (2007)
30. Badre, D., Poldrack, R.A., Paré-Blagoev, E.J., Insler, R.Z., Wagner, A.D.: Dissociable controlled retrieval and generalized selection mechanisms in ventrolateral prefrontal cortex. Neuron. **47**(6), 907–918 (2005)
31. Barch, D.M., Braver, T.S., Sabb, F.W., Noll, D.C.: Anterior cingulate and the monitoring of response conflict: evidence from an fMRI study of overt verb generation. J. Cogn. Neurosci. **12**(2), 298–309 (2000)
32. Beaty, R.E., Benedek, M., Silvia, P.J., Schacter, D.L.: Creative cognition and brain network dynamics. Trends Cogn. Sci. **20**(2), 87–95 (2016)
33. Boorman, E.D., Behrens, T.E., Woolrich, M.W., Rushworth, M.F.: How green is the grass on the other side? Frontopolar cortex and the evidence in favor of alternative courses of action. Neuron. **62**(5), 733–743 (2009)
34. Braver, T.S., Barch, D.M., Gray, J.R., Molfese, D.L., Snyder, A.: Anterior cingulate cortex and response conflict: effects of frequency, inhibition, and errors. Cereb. Cortex. **11**(9), 825–836 (2001)
35. Brunyé, T.T., Moran, J.M., Cantelon, J., Holmes, A., Eddy, M.D., Mahoney, C.R., Taylor, H.A.: Increasing breadth of semantic associations with left frontopolar direct current brain stimulation: a role for individual differences. Neuroreport. **26**(5), 296–301 (2015)
36. Bunge, S.A., Wendelken, C., Badre, D., Wagner, A.D.: Analogical reasoning and prefrontal cortex: evidence for separable retrieval and integration mechanisms. Cereb. Cortex. **15**(3), 239–249 (2005)
37. Cerruti, C., Schlaug, G.: Anodal transcranial direct current stimulation of the prefrontal cortex enhances complex verbal associative thought. J. Cogn. Neurosci. **21**(10), 1980–1987 (2009)
38. Chi, R.P., Snyder, A.W.: Brain stimulation enables the solution of an inherently difficult problem. Neurosci. Lett. **515**(2), 121–124 (2012)
39. Cho, S., Moody, T.D., Fernandino, L., Mumford, J.A., Poldrack, R.A., Cannon, T.D., et al.: Common and dissociable prefrontal loci associated with component mechanisms of analogical reasoning. Cereb. Cortex. **20**(3), 524–533 (2010)
40. Christoff, K., Gabrieli, J.D.: The frontopolar cortex and man cognition: evidence for a rostrocaudal hierarchical organization within the human prefrontal cortex. Psychobiology. **28**(2), 168–186 (2000)
41. Chrysikou, E.G., Hamilton, R.H., Coslett, H.B., Datta, A., Bikson, M., Thompson-Schill, S.L.: Noninvasive transcranial direct current stimulation over the left prefrontal cortex facilitates cognitive flexibility in tool use. Cogn. Neurosci. **4**(2), 81–89 (2013)
42. Colombo, B., Bartesaghi, N., Simonelli, L., Antonietti, A.: The combined effects of neurostimulation and priming on creative thinking. A preliminary tDCS study on dorsolateral prefrontal cortex. Front. Hum. Neurosci. **9**, 403 (2015)
43. Cromer, J.A., Roy, J.E., Miller, E.K.: Representation of multiple independent categories in the primate prefrontal cortex. Neuron. **66**(5), 796–807 (2010)
44. Geake, J.G., Hansen, P.C.: Neural correlates of intelligence as revealed by fMRI of fluid analogies. NeuroImage. **26**(2), 555–564 (2005)
45. Gilbert, S.J., Spengler, S., Simons, J.S., Steele, J.D., Lawrie, S.M., Frith, C.D., Burgess, P.W.: Functional specialization within rostral prefrontal cortex (area 10): a meta-analysis. J. Cogn. Neurosci. **18**(6), 932–948 (2006)
46. Goel, V., Eimontaite, I., Goel, A., Schindler, I.: Differential modulation of performance in insight and divergent thinking tasks with tDCS. J. Probl. Solving. **8**(1), 2 (2015)
47. Gonen-Yaacovi, G., de Souza, L.C., Levy, R., Urbanski, M., Josse, G., Volle, E.: Rostral and caudal prefrontal contribution to creativity: a meta-analysis of functional imaging data. Front. Hum. Neurosci. **7**, 465 (2013)
48. Green, A.E., Cohen, M.S., Kim, J.U., Gray, J.R.: An explicit cue improves creative analogical reasoning. Intelligence. **40**(6), 598–603 (2012)

49. Green, A.E., Cohen, M.S., Raab, H.A., Yedibalian, C.G., Gray, J.R.: Frontopolar activity and connectivity support dynamic conscious augmentation of creative state. Hum. Brain Mapp. **36**(3), 923–934 (2015)
50. Green, A.E., Kraemer, D.J., Fugelsang, J.A., Gray, J.R., Dunbar, K.N.: Neural correlates creativity in analogical reasoning. J. Exp. Psychol. Learn. Mem. Cogn. **38**(2), 264 (2012)
51. Hampshire, A., Thompson, R., Duncan, J., Owen, A.M.: Lateral prefrontal cortex subregions make dissociable contributions during fluid reasoning. Cereb. Cortex. **21**(1), 1–10 (2011)
52. Holyoak, K. J.: Analogy and relational reasoning. In: The Oxford Handbook of Thinking and Reasoning, pp. 234–259.
53. Ihne, J.L., Gallagher, N.M., Sullivan, M., Callicott, J.H., Green, A.E.: Is less more: does a prefrontal efficiency genotype actually confer better performance when working memory becomes difficult? Cortex. **74**, 79–95 (2016)
54. Miller, J., Berger, B., Sauseng, P.: Anodal transcranial direct current stimulation (tDCS) increases frontal–midline theta activity in the human EEG: a preliminary investigation of non-invasive stimulation. Neurosci. Lett. **588**, 114–119 (2015)
55. Sharon, M.: File: Little Horse on Wheels (Ancient Greek child's toy). Jpg. License: Creative Commons Attribution 2.0 Generic license. File:Little horse on wheels (Ancient greek child's Toy).jpg—Wikimedia Commons (2009)
56. Photo by Dana Barry
57. Photo provided by Dana Barry
58. Barry, D.M., Kanematsu, H., Kobayashi, T., Shimofuruya: Hiroshi. Multisensory science (Idea Bank). Sci. Teach. **66** (2003)
59. Barry, D.: Science Fair Projects. Published by Teacher Created Materials, Inc, Huntington Beach (2000)
60. Photo by Hideyuki Kanematsu
61. Barry, D.M., Kanematsu, H.: Develop Critical Thinking Skills, Solve a Mystery, Learn Science. U.S. Published by Tate Publishing (2007)
62. 12 Best innovative teaching strategies for modern pedagogy that can improve student engagement. The Scientific World (2019)
63. Innella, G., Rodgers, P.: The benefits of a convergence between art and engineering. In: High Tech and Innovation (2021)
64. Perales, F.J., Aróstegui, J.L.: The STEAM approach: implementation and educational, social, and economic consequences. Arts Educ. Policy Rev. **125**(2), 59–67 (2024). https://doi.org/10.1080/10632913.2021.1974997
65. Allina, B.: The development of STEAM educational policy to promote student creativity and social empowerment. Arts Educ. Policy Rev. **119**(2), 77–87 (2018)
66. Aróstegui, J.L.: Exploring the global decline of music education. Arts Educ. Policy Rev. **117**(2), 96–103 (2016)
67. Aróstegui, J.L., Kyakuwa, J.: Generalist or specialist music teachers? Lessons from two continents. Arts Educ. Policy Rev. **122**(1), 19–31 (2021)
68. Bencze, L., Reiss, M.J., Sharma, A., Weinstein, M.: CHAPTER SIX: STEM education as "Trojan Horse": deconstructed and reinvented for all. Counterpoints. **442**, 69–87 (2018)
69. Boytchev, P., Boytcheva, S.: Gamified evaluation in STEAM for higher education: a case study. Information. **11**(6), 316 (2020)
70. Colucci-Gray, L., Burnard, P., Gray, D., Cooke, C.: A Critical Review of STEAM (Science, Technology, Engineering, Arts, and Mathematics). Oxford Research Encyclopedia of Education (2019)
71. Herro, D., Quigley, C., Jacques, L.A.: Examining technology integration in middle school STEAM units. Technol. Pedagog. Educ. **27**(4), 485–498 (2018)
72. Hogan, J., Down, B.: A STEAM School using the Big Picture Education (BPE) design for learning and school—what an innovative STEM Education might look like. Int. J. Innov. Sci. Math. Educ. **23**(3) (2015)

73. Katz-Buonincontro, J.: Gathering STE (A) M: Policy, curricular, and programmatic developments in arts-based science, technology, engineering, and mathematics education Introduction to the special issue of Arts Education Policy Review: STEAM Focus. Arts Educ. Policy Rev. **119**(2), 73–76 (2018)
74. Perignat, E., Katz-Buonincontro, J.: STEAM in practice and research: an integrative literature review. Think. Skills Creat. **31**, 31–43 (2019)
75. Dell'Erba, M.: Policy Considerations for STEAM Education. Policy Brief. Education Commission of the States (2019)
76. Costantino, T.: STEAM by another name: transdisciplinary practice in art and design education. Arts Educ. Policy Rev. **119**(2), 100–106 (2018)
77. Patil, S., Vasu, V., Srinadh, K.V.S.: Advances and perspectives in collaborative robotics: a review of key technologies and emerging trends. Discov. Mech. Eng. (2023)
78. Zhou, C.: Fostering creative engineers: a key to face the complexity of engineering practice. Eur. J. Eng. Educ. **37**(14) (2012)
79. Selinger, C.: The creative engineer: what can you do to spark new ideas? IEEE Spectr. **41**(8), 47–49 (2004)
80. Taking a rigorous approach to creativity in civil engineering. Civil Engineering Magazine (2021)
81. Loots, E., van Bennekom, S.: Entrepreneurial firm growth in creative industries: fitting in … and standing out! Creative Ind. J. **16**(3), 383–405 (2023). https://doi.org/10.1080/17510694.2022.2025710
82. Abecassis-Moedas, C., Ben Mahmoud-Jouini, S., Dell' Era, C., Manceau, D., Verganti, R.: Key resources and internationalization modes of creative knowledge-intensive business services: the case of design consultancies. Creat. Innov. Manag. **21**(3), 315–331 (2012). https://doi.org/10.1111/j.1467-8691.2012.00646.x
83. Barney, J.B.: Measuring firm performance in a way that is consistent with strategic management theory. Acad. Manage. Discov. **6**(1), 5–7 (2020). https://doi.org/10.5465/amd.2018.0219
84. Bergamini, M., Van de Velde, W., Van Looy, B., Visscher, K.: Organizing artistic activities in a recurrent manner: (on the nature of) entrepreneurship in the performing arts. Creat. Innov. Manag. **27**(3), 319–334 (2018). https://doi.org/10.1111/caim.12240
85. Blackburn, R., Kitching, J., Hart, M., Brush, C., Ceru, D.: Growth Challenges for Small and Medium-Sized Enterprises: A UK-US Comparative Study. Report for HM Treasury and BERR. Kingston University and Babson College, London (2008)
86. Caniëls, M.C., Rietzschel, E.F.: Creativity: creativity and innovation under constraints. Creat. Innov. Manag. **24**(2), 184–196 (2015). https://doi.org/10.1111/caim.12123
87. Caves, R.E.: Creative Industries: Contracts Between Art and Commerce. Harvard University Press, Cambridge, MA (2000)
88. Chaston, I., Sadler-Smith, E.: Entrepreneurial cognition, entrepreneurial orientation and firm capability in the creative industries. Br. J. Manag. **23**(3), 415–432 (2012)
89. Cnossen, B., Loots, E., Witteloostuijn, A.: Individual motivation among entrepreneurs in the creative and cultural industries: a self-determination perspective. Creat. Innov. Manag. **28**(3), 389–402 (2019). https://doi.org/10.1111/caim.12315
90. de Klerk, S., Hodge, S.: The case of the creative accelerator. Creative Ind. J. **14**(2), 169–189 (2021). https://doi.org/10.1080/17510694.2020.1813480
91. Du, X., Zhang, H., Zhang, S., Zhang, A., Chen, B.: Creativity and leadership in the creative industry: a study from the perspective of social norms. Front. Psychol. **12** (2021)
92. Amabile, T.M., Conti, R., Coon, H., Herron, L.M.: Assessing the work environment for creativity. Acad. Manag. J. **39**, 1154–1184 (1996). https://doi.org/10.5465/256995
93. Baron, R.A., Tang, J.: The role of entrepreneurs in firm-level innovation: joint effects of positive effects, creativity, and environmental dynamism. J. Bus. Ventur. **26**, 49–60 (2011). https://doi.org/10.1016/j.jbusvent.2009.06.002
94. Bass, B.M.: From transactional to transformational leadership: learning to share the vision. Organ. Dyn. **18**, 19–31 (1990). https://doi.org/10.1016/0090-2616(90)90061-S

95. Blay, A.D., Gooden, E.S., Mellon, M.J., Stevens, D.E.: The usefulness of social norm theory in empirical business ethics research: a review and suggestions for future research. J. Bus. Ethics. **152**, 191–206 (2016). https://doi.org/10.1007/s10551-016-3286-4
96. Chen, M.H., Chang, Y.Y., Lin, Y.C.: Exploring creative entrepreneurs' happiness: cognitive style, guanxi and creativity. Int. Entrep. Manag. J. **14**, 1089–1110 (2018). https://doi.org/10.1007/s11365-017-0490-3
97. Chen, M.H., Chang, Y.Y., Lo, Y.H.: Creativity cognitive style, conflict, and career success for creative entrepreneurs. J. Bus. Res. **68**, 906–910 (2015). https://doi.org/10.1016/j.jbusres.2014.11.050

Chapter 2
Social Revolution Accompanying New Technology

Nobuyuki Ogawa

Abstract The rapid advancement of technology is ushering in a social revolution that transcends traditional boundaries. This revolution impacts various facets of society, including labor and employment, education and skills, social disparities, privacy and security, health and wellness, environment and sustainability, and cultural and social change. Key technological drivers include Artificial Intelligence (AI) and Machine Learning, the Internet of Things (IoT), Blockchain, Virtual Reality (VR) and Augmented Reality (AR), and Quantum Computing. These technologies reshape industries and societal norms, influencing productivity, remote collaboration, manufacturing, customer experience, and education. Case studies and prospects of Society 5.0 encompass the realization of Smart Cities, healthcare innovations, educational transformations, smart agriculture, e-government initiatives, smart retail, crowdsourcing, automated driving technology, digital twin applications, robotics, autonomous systems, and quantum computing. Metaverse and XR technology offer immense potential to enhance the industry by providing immersive experiences, facilitating collaboration in virtual spaces, conducting business activities, and bridging the gap between the real and virtual domains.

Keywords New technologies · Artificial intelligence · Machine learning · Internet of Things · Virtual reality · Society 5.0 · Social revolution

2.1 Introduction

Rapid advances in new technologies can potentially bring about significant changes in society. Below, we explore some aspects of the social revolution associated with the latest technologies.

N. Ogawa (✉)
National Institute of Technology, Gifu College, Gifu, Japan
e-mail: ogawa@gifu-nct.ac.jp

D. M. Barry, H. Kanematsu (eds.), *Applications of Metaverse and Virtual Reality to Creative Education and Industry*, Intelligent Systems Reference Library 267, https://doi.org/10.1007/978-981-96-3341-8_2

- ***Labor and Employment:*** [1–10]

New technologies will change the traditional labour market. Some jobs may be replaced by automation and robotization, but new jobs and skills will be created. Remote and flexible work arrangements will also become more common, changing the balance between employees' lives and work.

- ***Education and Skills:*** [11–17]

The proliferation of new technologies will change the nature of education: STEM education (science, technology, engineering, and mathematics) will become increasingly important, and digital literacy will be essential. At the same time, life-long learning will be emphasized, and individuals will be required to pursue personal growth continuously.

- ***Social disparities:*** [18–27]

The disparity between those more likely to benefit from new technologies and those less likely to do so may widen. There is concern that the gap in access to and use of digital technologies, known as the digital divide, will become more pronounced. Comprehensive efforts, including reforms in social policy and education, are needed.

- ***Privacy and Security:*** [28–35]

Advances in digital technology make managing personal information and data a critical issue. Increasing emphasis will be placed on privacy protection and security. Laws and regulations in various countries will be required to be in place, and technology companies and governments will be held accountable.

- ***Health & Wellness:*** [36–44]

Advances in medical technology and new health-related technologies will change how health services and health care are provided. Telemedicine and mobile health technologies will become more prevalent and improve access to health. At the same time, issues of personal health information management and medical ethics will be highlighted.

- ***Environment and Sustainability:*** [45–52]

New technologies must contribute to sustainability. Innovation in renewable energy, clean technology, and circular economy will be necessary. At the same time, the latest environmental issues may arise, such as the environmental impact of digitization and the increase in e-waste.

- ***Cultural and social change:*** [53–58]

New technologies also have a profound impact on culture and social structures. Media consumption patterns and communication methods will change, as will individual identities and relationships. At the same time, digital technologies are expected to promote expression and community formation, creating new cultural and social ties.

It is essential to take these aspects into account as we move forward with the introduction of new technologies and the transformation of society. Technological innovation and social development are closely linked and are fundamental to creating a sustainable and inclusive future.

The industry's outlook can change dramatically as new technologies evolve. Below are some key new technologies and their associated prospects for the industry.

1. ***Artificial Intelligence (AI) and Machine Learning:*** [59–68]

Advances in AI and machine learning will enable autonomous systems and processes to improve efficiency. For example, AI can be used in manufacturing for preventive maintenance and quality control to minimize production line downtime. AI chatbots and virtual assistants are expected to streamline customer service and improve customer satisfaction in the customer service sector.

2. ***Internet of Things (IoT):*** [69–76]

With the development of IoT, industrial sensors will become even more prevalent. They will monitor, control, and optimize products and processes, resulting in efficient resource utilization and quality control.

The concept of intelligent cities is evolving, making transportation, energy management, and public facilities more efficient and improving the sustainability of cities.

3. ***Blockchain:*** [77–84]

Blockchain technology provides reliable data sharing and transaction transparency, enabling supply chain management, contract automation, and enhanced data security.

In the financial industry, blockchain-based smart contracts and decentralized exchanges will improve the efficiency and security of transactions.

4. ***Virtual Reality (VR) and Augmented Reality (AR):*** [85–94]

The proliferation of VR and AR will make remote working, training, and product design more immersive and compelling. This will foster collaboration among global teams, increasing creativity and productivity.

AR is expected to streamline maintenance and troubleshooting in the manufacturing industry, improving worker skills and productivity.

5. ***Quantum Computing:*** [95–104]

The rise of quantum computing will significantly increase the speed of solving complex problems and analyzing data, accelerating discoveries and innovations in medicine and materials science. It enables new approaches to supply chain management, optimization problems, and other issues that conventional computers cannot solve.

These new technologies have the potential to revolutionize industry and improve society's sustainability and efficiency. Over the coming decades, their evolution will be integral to industrial development.

In addition, integrating new technologies and the metaverse opens many possibilities for industry. Some of the main points are listed below.

- ***Virtual Product Development and Design:***

Using virtual space within the Metaverse dramatically changes the product development process. Product design and prototyping will be faster and more efficient, allowing for real-time prototyping and modification. AR and VR can also experience product design and functionality in a realistic environment. This improves product quality and user experience and enhances market competitiveness.

- ***Remote Work and Collaboration:***

By utilizing virtual offices and meeting rooms in the Metaverse, remote work can be achieved beyond geographical constraints. Employees can communicate and collaborate on projects and tasks in real-time and within the virtual space. Such an environment allows for the best use of human resources, regardless of geographic location or time zone. In addition, the combination of collaboration tools and digital whiteboards facilitates idea-sharing and brainstorming.

- ***Manufacturing and Maintenance:***

Metaverse and AR can help workers efficiently assemble and maintain products in manufacturing. AR headsets and wearable devices can display work and troubleshooting procedures in real-time, allowing workers to respond quickly.

Furthermore, combined with AI, it can detect product abnormalities and failures and perform preventive maintenance activities, reducing production line stoppages and maintenance costs.

- ***Customer Experience and Sales:***

Metaverse-based shopping experiences and virtual try-on services will expand. Customers can now experience actual stores and products from their homes and make better purchasing decisions.

Combined with AI, analyzing customer preferences and purchase history and providing individually optimized products and services is possible. It will provide services to customers. This will increase customer satisfaction and loyalty, leading to increased sales.

- ***Training and Education:***

Simulation and training programs utilizing the Metaverse will expand. Employees can receive hands-on training in a risk-free virtual environment to acquire skills and improve operational efficiency. Students and pupils can learn in a virtual space and gain a deeper understanding through realistic experiences. This improves the quality of education and enhances learning.

The convergence of new technologies and the metaverse will enable more efficient and innovative business models and services than ever before. The industry must proactively embrace and leverage these technologies to enhance competitiveness and achieve sustainable growth.

The convergence of metaverse and artificial intelligence (AI) has the potential to revolutionize the industry. Metaverse is a digital space constructed through virtual reality (VR) and augmented reality (AR), providing an environment where the real and digital worlds merge. AI is a system that can extract patterns from data and solve problems autonomously using machine learning and deep learning techniques. The combination of these two technologies offers the following industrial possibilities.

1. ***Increased Productivity:***

The convergence of metaverse and AI can potentially increase productivity in the manufacturing and production industries. For example, AI-powered robots and automation systems can be controlled in the metaverse to optimize manufacturing lines and factory operations. Workers can be trained in a virtual reality environment to analyze data in real-time and perform tasks efficiently.

2. ***Facilitating remote work:***

Metaverse and AI could further develop the remote work environment. Virtual offices and meeting rooms can be built on the metaverse, and AI can help facilitate meetings and organize materials. This allows employees to collaborate without geographical restrictions and allows for a more flexible work style.

3. ***Improved Customer Experience:***

Metaverse-powered AI can improve the customer experience. For example, AI-powered virtual assistants can answer customer questions in real-time and provide information about products and services. They could also leverage AR to enable customers to experience products and services virtually.

4. ***Improved data analysis and forecasting accuracy:***

AI analyzes large amounts of data collected on the metaverse and makes trends and forecasts to improve decision-making accuracy. This allows companies to understand market trends and customer needs more accurately, which can be used to develop strategies and optimize business processes.

5. ***Creation of new industries:***

The convergence of metaverse and AI may also facilitate the creation of new industries. Examples include events, experiential services in virtual spaces, and the development of virtual personalities through AI. This will create new business models and services and promote industry diversification.

As described above, the convergence of metaverse and AI has the potential to impact the industry profoundly. The use of these technologies will bring a variety of benefits, including increased productivity, improved customer experience, and the creation of new business opportunities.

2.2 What Is Society 5.0?

This chapter provides an overview of Society 5.0 (an innovative and sustainable society based on the convergence of digital technology and physical space) proposed by the Japanese government and its key features. It explores trends toward its implementation and the strengthening of local communities. Each feature and why it is a critical element of Society 5.0 will be explained.

2.2.1 Background

To understand the background of Society 5.0, it is essential to consider the history of human social evolution. Technological advances and social changes have greatly influenced the development of human civilization, and it is necessary to follow the process from Society 1.0 to Society 4.0 and understand how Society 5.0 emerged from this process.

- ***Society 1.0: Hunter-Gatherer Society***

Society 1.0 refers to an early form of human civilization, the hunter-gatherer society. In this period, people lived off nature's bounty and earned their livelihood through hunting and gathering. Humans lived in groups and were organized into tribes and clans. Language and tool use evolved, forming the foundations of culture and society.

- ***Society 2.0: Agricultural Revolution and Urbanization***

Society 2.0 was the era of the Agricultural Revolution when agricultural technology development led to increased productivity. This led to a rapid increase in population and the formation of cities. The exchange of people and goods between urban and rural areas became more active, and commerce developed. This led to changes in social organization and economic systems and new social strata.

- ***Society 3.0: Industrial Revolution and Industrialization***

Society 3.0 was the era of the Industrial Revolution when the development of mechanical technology led to industrialization. The invention of the steam engine and mechanical tools made production processes more efficient and enabled mass Production. This led to rapid industrial development and accelerated urbanization. Working and capitalist classes were formed, and the economic and political structures changed dramatically.

- ***Society 4.0: The Digital Revolution and the Information Society***

Society 4.0 is characterized by the rapid evolution of digital technology and the spread of the Internet. The information society has been formed, and the distribution of information has expanded rapidly. The development of computers and the Internet has dramatically improved information and communication, creating a "global village" that connects the world. Digital technology has influenced every aspect of the economy and society, creating new industries and business models. Society is in the final stages of Society 4.0 and is in a transitional stage of moving to Society 5.0 shortly.

- ***Society 5.0: The Fusion of Digital and Physical***

Based on this history of social evolution, Society 5.0 has been proposed, which will make further social evolution possible through the fusion of digital technology and physical space. While building on the progress of digitalization up to Society 4.0, the fusion of digital technology and physical space has become critical. Physical space and digital technology must be integrated to make people's lives and industries more innovative and sustainable. This will improve people's quality of life and the sustainable development of society.

In Society 5.0, all elements of society will be constructed as a digital twin in cyberspace, reconfigured in terms of institutions, business design, and urban and regional development, and then reflected in physical space to transform society. By incorporating the value of human-centeredness into such a new process, society will change flexibly and agilely for the better, with each citizen and citizen of the world as a central figure in the decision-making arena.

2.3 Current Society (Society 4.0)

To understand the society of the near future (Society 5.0), it is essential to understand the current status and challenges of the current society (Society 4.0).

Society 4.0 is characterized by the rapid evolution of digital technology and the widespread use of the Internet. This era has facilitated the dissemination and sharing of information and given rise to new business models and services. Below is an overview of Society 4.0 and a detailed description of its main characteristics.

2.3.1 The Rise of the Digital Revolution

The rapid evolution of digital technology characterizes the advent of Society 4.0. Technologies such as computers, the Internet, and mobile devices have evolved daily and have had a significant impact on the lives of people around the world. Following are some of the critical features of this era.

– ***Information dissemination:***

The Internet has made collecting, sharing, and communicating information easier. Information is now available in various formats, including text, images, and video, and people worldwide can access it in real time.

– ***Increased connectivity:***

Technologies such as smartphones, tablets, and mobile internet allow people to stay connected, communicate, and exchange information in real-time. This will enable people worldwide to connect instantly, forming communities and global networks.

– ***Increased digitization:***

The shift from paper documents and media to digital formats accelerated as digitization progressed. This has facilitated data management and processing and streamlined information retrieval and analysis.

– ***Rise of new industries:***

New industries that leverage technologies like the Internet, social media, and cloud computing have emerged. This has created new business models and services in entertainment, telecommunications, advertising, and e-commerce.

2.3.2 Development of the Information Society

Society 4.0 has brought about the development of an information society. The information society is a form of society in which information and knowledge are at the center of the economy and society, and the distribution and use of information are significant factors in economic growth and social evolution. The following are the main characteristics of the information society.

– ***Formation of the Knowledge Economy:***

Knowledge and information became the primary resources of the economy, creating an economic structure in which the creation and utilization of knowledge was the key to economic growth. Companies and organizations seek a competitive edge by focusing on knowledge creation, maintenance, and sharing. Educational Transformation: In the information society, the forms and roles of education have changed dramatically. Information and communication technologies (ICT) have been introduced into the educational process, transforming learning. New forms of education, such as lifelong learning and online education, have emerged.

– ***Diffusion of access to information:***

The widespread use of the Internet has dramatically increased access to information. This has facilitated the sharing and learning of information and improved the ability of individuals and organizations to use information.

2.3.3 Issues for Society 4.0

The digital revolution and the progress of the information society in Society 4.0 present several challenges and problems. Some of them are described below.

- ***Digital Divide:***

The digital revolution can create disparities in access to and use of digital technologies. Due to economic factors and geography, some people and regions lack access to digital technologies. This may widen the digital divide and increase social inequality.

- ***Privacy and security concerns:***

With the development of the information society, personal privacy and data security is a growing concern. Digital technology increases the risk of unauthorized collection and misuse of personal information. Security issues such as cyber-attacks and data leaks are also serious concerns.

- ***Employment Change and Labor Market Instability:***

There is concern that the digital revolution will result in some jobs being replaced by automation and robotization. This could result in the loss of workplaces for some workers or a skills mismatch. Lack of skills and education to adapt to new technologies will also contribute to employment changes and labor market instability.

- ***Information Overload and Information Disparity:***

In the information society, large amounts of information are instantly accessible, but on the other hand, problems of information overload and credibility arise. Information overload can make it impossible to keep up with the capacity to process information. In addition, issues related to the authenticity and quality of information are becoming more serious, and information gaps may occur.

- ***Digital Dependence and Social Isolation:***

With the spread of digital technology, people's lives and social activities tend to be online. At the same time, there are concerns that digital dependency and excessive use of social media may lead to a loss of connection with the real world, an increase in social isolation, and psychological health problems.

These challenges and issues will likely become more severe as Society 4.0 progresses. For society to address these challenges, a wide range of initiatives will be needed, including the development of comprehensive policies and regulations, the ethical use of technology, the dissemination of education and skills, and the promotion of digital inclusion.

2.4 Transformation from Society 4.0 to Society 5.0

The following approaches are essential to addressing the challenges and issues of the digital revolution and information society in Society 4.0 and facilitating its development into Society 5.0.

– ***Bridging the Digital Divide:***

Bridging the digital divide requires narrowing the gap in access to and use of digital technologies. Comprehensive efforts are needed, including collaboration among governments, businesses, NGOs, and others to improve infrastructure. Digital education and access should be promoted to low-income groups and geographically isolated areas.

– ***Enhanced privacy and security:***

Addressing privacy and security issues requires enhanced legal regulations and technical measures. Proper management and protection of personal data, enhanced security measures, and end-to-end encryption are essential. Ethical data use, and transparency are also necessary.

– ***Skills Development and Retraining:***

Providing opportunities for skill upgrading and retraining is essential to keep pace with changes in employment. It is necessary to support the development of digital expertise and literacy to help workers adapt to new occupations and technologies.

– ***Developing a reliable information environment:***

Addressing information overload and credibility issues requires the development of a credible information environment. It is essential to ensure the credibility of digital and social media, prevent the spread of fake news and hoaxes, and improve information quality and source transparency.

– ***Digital Technology and Innovation:***

Promoting the use and innovation of digital technology and its deployment to Society 5.0 is essential. It is necessary to promote digitalization in industry and the public sector, create new business models and services, and contribute to a sustainable society.

– ***Building a human-centered society:***

An essential part of the development of Society 5.0 is constructing a human-centered society. The development of digital technology and information society should aim to improve the quality of people's lives and culture. Promoting the use of technology and the transformation of society, considering people's needs and values, is essential.

Through these efforts, it will be possible to resolve issues and problems in Society 4.0 and expand to Society 5.0. A comprehensive approach and cooperation among diverse stakeholders are essential for a sustainable society.

Introduce and utilize specific technologies. Some of these are detailed below. This will help solve the issues and problems of the digital revolution and information society in Society 4.0 and facilitate the development of Society 5.0.

– ***Artificial Intelligence (AI) and Machine Learning:***

Advances in AI and machine learning will enable data analysis and decision-making automation.

This will result in more efficient business processing and problem-solving, increasing productivity.

Introducing AI-based predictive analysis and optimization algorithms will ensure efficient resource use and service optimization. In addition, introducing AI self-learning functions will create a system that can continually adapt to the latest data and conditions.

– ***Blockchain Technology:***

Blockchain technology ensures reliable data sharing and transaction transparency. It Reduces the risk of data tampering and unauthorized access, improving reliability and security.

Using smart contracts and decentralized applications will eliminate centralized systems and processes and improve reliability and efficiency. In addition, blockchain is expected to enable digital identity management and supply chain traceability.

– ***Internet of Things (IoT):***

The proliferation of IoT technology will connect the physical and digital worlds in real-time. Sensor data collection and analysis will enable efficient resource management and product monitoring.

Using the IoT to create smart cities and homes will improve sustainability. Energy efficiency, optimized transportation, and enhanced living environments will be realized.

– ***Augmented Reality (AR) and Virtual Reality (VR):***

AR and VR technologies enhance remote working, training, and collaboration. By providing a realistic experience, communication and cooperation can transcend physical distance.

For example, using AR and VR in the manufacturing and medical sectors will improve training and surgical assistance for workers and medical personnel, increasing efficiency and safety.

Introducing and utilizing these specific technologies can solve issues and problems in Society 4.0 and accelerate the development of Society 5.0. However, a comprehensive approach and addressing social issues are necessary to link these technologies not only to technology but also to improving people's lives and the quality of society.

2.5 Society 5.0 Case Studies and Prospects

Society 5.0 is a concept that seeks innovation and sustainability with advanced technologies to realize a human-centered society. Below are some specific examples and perspectives of Society 5.0.

– ***Realization of Smart Cities:***

The realization of a smart city requires multiple technologies and initiatives. Below are some of the key elements to realize a smart city.

1. Infrastructure Development: To support smart cities, fast and reliable telecommunications infrastructure is needed. This includes laying optical fibre and building 5G networks.
2. Use of sensor technology: Sensors will be installed throughout the city to collect real-time data on traffic flow, environmental conditions, energy consumption, etc. This will enable efficient management of city functions and resource usage.
3. Utilization of Big Data and IoT: Collected data will be analyzed to understand the challenges and needs of the city; IoT devices and big data analysis technologies will be used to predict traffic congestion and efficient use of energy.
4. Optimize transportation systems: We make transportation systems more efficient, including self-driving cars and electrification of public transportation, to reduce traffic and congestion. Also important is the development of infrastructure to improve traffic safety.
5. Optimize energy management: Introduce renewable energy sources and establish smart grids to achieve efficient energy use and supply stability. This will also include energy demand forecasting and peak hour control.
6. Citizen Participation and Community Formation: Citizen and community participation will be encouraged, and feedback will be collected on city policies and plans. Also important are community centers, events, and other efforts to increase community cohesion.
7. Digital service delivery: Leverage digital services provided by government and private companies to make life easier for citizens. Examples include using smartphone apps to access public services and simplify online procedures.
8. Sustainable Urban Development: Reducing environmental impact and efficient resource use are important to realizing smart cities. Efforts must be made to promote sustainable urban development and reduce the burden on the global environment.

These factors will be planned and implemented to achieve a smart city. The goal is to combine various factors, including technological advances, citizen participation, and cooperation between government and the private sector, to achieve more sustainable and efficient cities.

– ***Health Care Innovations:***

Healthcare innovation refers to how medical technology and information technology advances transform how healthcare services are provided and managed. Below are some key elements of healthcare innovation

1. Telemedicine and Telehealth: Telemedicine advancements have allowed patients to receive video calls and online consultations with physicians from home or remote locations. Telemedicine improves the accessibility of healthcare services and meets demand, especially in geographically remote areas and among the elderly.
2. Wearable Devices and Healthcare Apps: The proliferation of wearable devices such as smartwatches, fitness trackers, and healthcare apps enables individuals to monitor their health status in real-time and record and analyze data. This will promote increased health awareness and early detection of diseases.
3. Artificial Intelligence and Medical Diagnosis: Artificial intelligence (AI) technology advances have advanced medical diagnostics, including analysis of medical images and pathological diagnosis; AI can find patterns in large amounts of medical data to assist physicians in diagnosing and optimizing treatment plans.
4. Gene Therapy and Personalized Medicine: Advances in gene therapy and genetic testing technologies have enabled personalized medicine based on a patient's genetic information. This allows for the assessment of a patient's disease risk and treatment efficacy at the genetic level and the development of more effective treatment plans.
5. Health Data Integration and Utilization: Efforts are underway to integrate health data held by medical institutions, insurance companies, and research institutes to build a large-scale health database. This will expand the scope of utilization of health information, such as disease prevalence and evaluation of treatment effectiveness.
6. Virtual Care and Health Coaching: Virtual and augmented reality (VR/AR) technologies are now used to provide virtual care services such as rehabilitation, stress management, and health education. In addition, health coaching applications promote personal health goal-setting and healthy behaviors.

These innovative technologies and initiatives are transforming the way traditional medical services and healthcare are delivered in the healthcare sector to provide more efficient and personalized care.

– ***Educational Transformation:***

The transformation of education in Society 5.0 aims to provide education that meets individual abilities and needs through technological innovation, thereby realizing a human-centered society. Below are some elements of this transformation.

1. Promoting Personalized Learning: Society 5.0 will leverage AI and big data analytics technologies to provide customized learning experiences tailored to students' individual learning styles and progress. Students can learn more effectively according to their interests and abilities.

2. Creating a Flexible Learning Environment: The development of online educational platforms and distance learning systems will enable learning that is not restricted by time or location. Students and scholars can learn independently while working across geographic boundaries.
3. Developing Digital Skills: In Society 5.0, digital skills are becoming increasingly important as digital technology develops. Education will focus on developing digital skills such as programming, content creation, and information literacy to help students and faculty adapt to the digital world.
4. Fostering Collaboration and Collaborative Learning: In an increasingly globalized world, it is vital to interact and collaborate with people from different cultures and backgrounds. Education provides students and faculty opportunities for collaborative learning and cross-cultural understanding through online collaboration tools and international exchange programs.
5. Promoting Life-Long Learning: Society 5.0 will emphasize lifelong learning as people's life stages and careers change. Educational institutions and businesses will work together to provide job training and skill development opportunities to support individual growth and career development.
6. Use and Improvement of Educational Data: Educational institutions and governments collect data to improve educational systems and programs. Through data analysis, we evaluate educational effectiveness and student progress and pursue the delivery of education that meets individual needs.

Combining these elements will drive the transformation of education in Society 5.0. To realize a human-centered society, an educational system is expected to be created that optimizes the relationship between technology and people and promotes individual growth and social development. It is expected.

– ***Smart Agriculture:***

Smart agriculture is an effort to improve productivity and achieve sustainable agriculture with the use of the latest technology in agriculture. Below are some specific features and benefits of intelligent agriculture.

1. Use of sensor technology: Soil, climate, water quality, and other data are collected in real-time, and sensor data are analyzed to determine crop growth and environmental conditions. This allows for proper cultivation management and efficient resource use.
2. Drones and satellite imagery are used to monitor large areas of farmland efficiently. This enables early detection of abnormal growth conditions, pest damage, etc., and prompt response.
3. Automation and Robot Technology: Automating agricultural work and introducing robot technology can help alleviate labor shortages and improve work efficiency. For example, automated guided vehicles and harvesting robots will be utilized.
4. Optimization of water resource management: IoT sensors will control irrigation systems for more efficient water use. Water resources can be conserved by

monitoring groundwater and surface water levels and providing the right amount of water at the right time.
5. Data analysis and forecasting: Historical data, weather data, and market trends are analyzed to support decision-making for planting and shipping plans. Demand and price forecasts help you make efficient business plans.
6. Promote sustainable agriculture: Reduce chemical fertilizers and pesticides and promote organic and environmentally friendly cultivation methods. This is expected to reduce contamination of soil and water sources and improve food safety and quality.

Smart farming improves productivity and the use of resources more efficiently. It reduces environmental impact, making it an essential step toward sustainable agriculture.

– ***E-Government:***

E-Government refers to efforts by government agencies to provide administrative services and information, improve the efficiency of administrative procedures, and promote citizen participation through information and communications technology (ICT). Some of the key features and benefits of e-government are listed below.

1. Provision of online services: Government agencies offer various administrative services through their websites and applications. For example, tax payments, social insurance applications, and licensing procedures can be made online.
2. Transparency and openness of information: Information on government activities, budgets, and policies will be made available online, making it easier for citizens to understand government activities. This increases government transparency and enhances citizen trust.
3. Streamlining Administrative Procedures: e-Government simplifies and streamlines administrative procedures using digitized processes and online forms. This allows citizens and businesses to complete procedures faster and more smoothly.
4. Facilitating citizen participation: By providing a forum for online exchanging of opinions and discussion, citizens can more easily participate in policymaking. Online voting and citizen surveys can also be conducted, enabling policymaking that reflects citizens' views.
5. Data utilization and analysis: In e-government, data collected from citizens and businesses are analyzed and used for policymaking and evaluation. This will lead to more effective policy implementation and improved public services.
6. Information sharing among regions: Utilizing the e-government framework, information will be shared among local governments and national and local governments. This will promote inter-regional coordination and cooperation, contributing to regional development.

The introduction of e-government improves the convenience of government services and increases government administration's efficiency and transparency, thereby improving citizens' lives and the business environment.

– ***Smart Retail:***

Smart retailing refers to the process of transforming the retail industry through the widespread use of digital technology and the Internet. Some of the key characteristics and initiatives of smart retailing are listed below.

1. Digitized store experience: Smart retailing provides an engaging in-store experience for customers by introducing interactive displays, touch-screen terminals, and augmented reality (AR) technologies that leverage digital technology within the store. This provides product information and detailed customization options to keep customers engaged.
2. Deploying an Omni-Channel Strategy: Smart retail emphasizes an omnichannel strategy that seamlessly integrates online shopping with physical stores. Customers will have multiple purchasing channels at their disposal, including the ability to purchase products through a website or mobile app, pick them up in-store, and return or exchange them.
3. Leveraging Customer Data: Smart Retail collects and analyzes data on customer behavior and purchase history to understand customer preferences and needs and develop individually tailored marketing strategies. By offering targeted promotions and special offers to customers, the company seeks to increase their willingness to purchase and build loyalty.
4. Automation and Robotics Deployment: Smart retailing will improve operational efficiency in inventory management, stocktaking, and logistics management through the implementation of automation technology and robotics. Automated processes reduce human error, shorten work hours, and lower store operating costs.
5. Artificial Intelligence and Personalization: Smart retailing leverages artificial intelligence (AI) technology to provide personalized product recommendations and customer support; AI-powered personalized service and product recommendations help improve customer satisfaction and the buying experience.
6. Security and Privacy: Smart retail emphasizes technology and policies to ensure the security of customers' personal information and transaction data. It is important to collect and use data without compromising customer trust.

Smart retailing will improve the overall competitiveness of the retail industry by enhancing the customer experience and improving operational efficiency. Advancements in technology are expected to result in a more innovative and customer-centric retail experience.

– ***Crowdsourcing:***

Crowdsourcing is a form of labor market in which tasks and projects are shared and executed by multiple individuals and organizations via the Internet. Below are some of the features and advantages of crowdsourcing

1. Flexible Labor Forms: Crowdsourcing is suitable for freelance and side activities because tasks can be performed without any restrictions on location or time. It allows for flexible work that fits your personal lifestyle and life stage.

2. Leveraging the global workforce: Crowdsourcing platforms via the Internet allow workers worldwide to participate in tasks. This enables the use of a diverse workforce that transcends geographical constraints.
3. Increased cost efficiency: Crowdsourcing enables efficient talent management by selecting the right workforce for the required task and paying them only when needed. In addition, labor costs may be lower than traditional forms of employment.
4. Ensure Scalability: Crowdsourcing can be flexible enough to handle large projects or temporary tasks, which can help companies and organizations recruit talent and scale projects during rapid growth.
5. Leveraging expertise and skills: Crowdsourcing platforms are populated with workers with expertise and skills in various fields. This allows you to select the right people for tasks that require specific expertise and skills to ensure high-quality deliverables.
6. Fostering Innovation: Crowdsourcing is where diverse talent and ideas can combine to generate innovation and diversify problem-solving approaches. Companies and organizations can leverage outside knowledge and experience to drive new ideas and projects.

These advantages have made crowdsourcing a vital business model and form of labor for many industries and companies. However, challenges exist, such as imbalances in working conditions and compensation and privacy concerns. Therefore, proper management and consideration are required when utilizing crowdsourcing.

– ***Automated Driving Technology:***

Automated driving technology refers to the ability of a vehicle to travel autonomously without the intervention of a human driver. Below are some of the key features and benefits of automated driving technology

1. Increased Safety: Automated driving technology utilizes sensors, cameras, radar, and other devices to detect surrounding conditions in real-time and make safe driving decisions. This is expected to reduce accidents caused by human error or inattention while driving.
2. Traffic Flow Optimization: Automated vehicles can understand traffic conditions in real time through communication with other vehicles and infrastructure and efficiently control routes and speeds. This reduces traffic congestion and accidents and can smooth traffic flow.
3. Increased Comfort and Convenience: Automated vehicles allow drivers to relax or engage in other activities since they do not need to concentrate on the driving controls. They can also be used as a means of transportation for people with difficulty driving, such as the elderly and people with disabilities.
4. Environmental Considerations: Automated vehicles are expected to reduce fuel consumption and emissions through more efficient engines and traffic. They can also reduce traffic and carbon dioxide emissions by improving energy efficiency and optimizing traffic.

5. Creation of new business models: The widespread use of automated driving technology could lead to new business models and services. Examples include self-driving car-sharing services, delivery services, and mobility services.
6. Reduction of traffic accidents: Automated driving technology is expected to reduce traffic accidents caused by human error or sudden lack of reaction. Advanced control by sensors and algorithms improves safety and reduces the risk of traffic accidents.

With these advantages, automated driving technology is expected to improve traffic safety, efficiency, and convenience and create new business opportunities. However, many technical, legal, and ethical issues must be overcome.

- ***Digital Twin:***

A digital twin is a virtual model or simulation of a real-world physical object or process in a digital space. Below are some of the features and advantages of a digital twin.

1. Real-time visualization and analysis: Because the digital twin models physical objects and processes in real time, their state and behavior can be visualized and analyzed in real time. This allows you to understand the physical system's state and make appropriate decisions.
2. Predictive Analysis and Scenario Validation: Because the digital twin models the behavior of physical objects and processes, it is possible to predict future states and behavior and validate scenarios. This makes it easier to identify risks and develop countermeasures for future situations.
3. Remote Monitoring and Control: Because the digital twin models physical objects and processes in real-time, their condition can be remotely monitored and controlled as needed. Remote manipulation and intervention streamline the operation and management of physical systems.
4. Efficient Design and Development: The digital twin streamlines the design and development process of products and systems by providing a digital model of the physical object or process. Through simulation and testing, the digital twin can improve Product quality and performance.
5. Ease of modification and improvement: Because the digital twin is a digital model of a physical object or process, it is relatively easy to modify and improve. The digital twin can be used to test different scenarios and conditions and to improve products and systems.

The digital twin is an innovative technology that links the physical and digital worlds to enable real-time monitoring, analysis, prediction, and control. The digital twin is becoming increasingly important with the digitalization of industries and products.

- ***Robotics and Autonomous Systems:***

Robotics and autonomous systems are critical elements of Society 5.0. An overview is provided below:

Robotics and Autonomous Systems

Robotics is a field that combines mechanical engineering, electronics, computer science, and other technologies to design, manufacture, and control mechanical systems that perform tasks for humans. Autonomous systems can recognize their environment and perform tasks without human intervention.

In Society 5.0, robotics and autonomous systems will play the following roles.

1. Increased efficiency in the industry: Autonomous robots and systems will make production processes and manufacturing operations more efficient. For example, automated warehouse systems and production lines increase operations' speed and accuracy.
2. Innovations in the service industry: Robots and autonomous systems, including hotels, restaurants, and medical facilities, are also used in the service industry. For example, robotic front desks in hotels and automated food preparation systems in restaurants can help streamline service delivery and improve the customer experience.
3. Improved Transportation Systems: Autonomous transportation systems, such as self-driving cars and drones, can help prevent traffic accidents and optimize traffic flow, improving traffic safety and efficiency.
4. Disaster Response: Robots and drones are used for rescue operations at disaster sites and surveying affected areas. Autonomous systems work in environments where humans may be at risk.
5. Aging Society: Robotic technology can help in caregiving and assisted living in an aging society. Caregiving robots and automated life support systems can help improve the quality of life for older people.
6. Environmental Protection: Autonomous systems can help monitor the environment and efficiently use renewable energy. For example, autonomous drones are used for early detection and monitoring of forest fires.

The evolution of robotics and autonomous systems brings revolutionary changes to industry and society and is a critical element in realizing Society 5.0.

– ***Quantum computing:***

Quantum computing is a method of computing that processes information using the principles of quantum mechanics. Unlike conventional computers, it uses quantum bits (qubits), or quantum states, to represent information and takes advantage of properties such as quantum parallelism and quantum interference to perform calculations. Some of the features and advantages of quantum computing are listed below.

1. Quantum Parallelism: Quantum computers can perform multiple calculations simultaneously using the principle of superposition of quantum bits. This allows for faster computations than possible with conventional computers.
2. Quantum Interference: Quantum computers can exploit interference effects between qubits to increase the accuracy of calculations. Complex problems that conventional computers cannot solve can be addressed by taking advantage of quantum interference.

3. Exponential Performance Improvements: Quantum computers can improve performance exponentially by using quantum parallelism and interferometry. This allows them to solve complex and optimization problems efficiently.
4. Improved cryptanalysis and encryption techniques: Quantum computers have the potential to decode conventional cryptographic techniques such as prime factorization efficiently. On the other hand, quantum cryptography could lead to more secure encryption techniques.
5. Applications in Material Science and Drug Design: Quantum computers can simulate the behavior of molecules and materials, and thus, they have potential applications in areas such as material science and drug design. Quantum chemical simulations enable advanced analyses that are impossible with conventional computers.

Although many challenges remain in the development of quantum computing, its potential is enormous, and it is expected to bring about innovations in various fields such as science, medicine, security, communications, and finance.

Quantum computing is a method that uses the principles of quantum mechanics to process information. The principles of quantum computing contain several vital concepts.

1. Quantum bits (qubit): The fundamental element of quantum computing is the qubit. A qubit is represented based on the principles of quantum mechanics and can have a superposition state of zeros and ones. Unlike conventional bits, qubits can represent both states simultaneously, which is called superpositions.
2. Quantum superposition: When a qubit is in the 0 and 1 superposition state, it can have both 0 and 1 states at the same time. This property allows a quantum computer to perform multiple calculations simultaneously.
3. Quantum Interference: In a quantum computer, qubits can exhibit interference effects with each other. This is when the states of the qubits are in a superposition, and the interference between qubits passing through different paths can enhance or cancel out a particular state.
4. Quantum Entanglement: In a quantum computer, multiple qubits can be in a particular correlated state called entanglement. When an entangled qubit's state changes, the other qubits' states change simultaneously.
5. Quantum operations: Quantum computers use the superposition state of qubits, interference effects, and entanglement to perform quantum operations called quantum gates. A quantum gate is an operation that changes the state of a qubit and is used to perform quantum algorithms.
6. Quantum Measurement: After a quantum computer has calculated, it performs a quantum measurement to obtain the result. The state of the qubit is observed, and the result is converted into classical information.

Using these principles of quantum mechanics, quantum computers can perform more advanced computations than conventional computers. Research on quantum computing is ongoing, and its expanding range of applications is expected to solve new scientific and technological challenges and bring about innovation.

These examples (Realization of Smart Cities, Health Care Innovations, Educational Transformation, Smart Agriculture, E-government, Smart Retail, Crowdsourcing, Automated Driving Technology, Digital Twin, Robotics and Autonomous Systems, and Quantum computing) illustrate concrete efforts in various fields to realize Society 5.0. These initiatives take an approach that combines advanced technologies with solutions to social issues to realize a human-centered society.

2.6 Conclusion

2.6.1 Fusion of Cyber and Physical Space

From a Society 5.0 perspective, the convergence of cyberspace and physical space is one of the key elements for realizing a human-centered society. Society 5.0 is a concept that seeks to revolutionize people's lives and economic activities through the convergence of cyberspace and physical space to achieve a sustainable society.

Society 5.0 is a concept for a human-centered society, and XR (Extended Reality) is the collective term for

Augmented Reality (AR), Virtual Reality (VR), and Mixed Reality (MR). The combination of Society 5.0 and XR is how the next generation of technology will impact people's lives and businesses. Below are a few key points about the relationship between Society 5.0 and XR.

– ***Fusion of real and digital:***

XR technology can merge the real world with the digital world. For example, using AR to superimpose digital information on the natural environment can create rich experiences and interactions. From a Society 5.0 perspective, this fusion of real and digital has the potential to revolutionize people's lives and businesses, providing a more efficient and comfortable environment.

– ***Digital Twin and Simulation:***

A digital twin is a technology that reproduces objects and processes in physical space in a digital space; by utilizing XR technology, a digital twin can be visualized in a realistic environment and manipulated in real time. This facilitates experiments and simulations in physical space and improves the efficiency of product development and design processes.

– ***Education and Training:***

XR technology offers innovative methods in education and training. Simulation and training programs using virtual reality provide training in a safe and realistic environment and help develop practical skills. From a Society 5.0 perspective, education and training using XR can promote the development of individual competencies and skills and contribute to human resource development.

– ***Remote Collaboration:***

XR technology is also used as a tool to enhance remote collaboration. It enables people in remote locations to collaborate in real-time by sharing a virtual space. From a Society 5.0 perspective, this approach to remote collaboration overcomes geographical limitations and promotes collaboration and cooperation among people.

The combination of Society 5.0 and XR suggests the changes the next generation of technology will bring to society. The convergence of real and digital educational innovations and enhanced remote collaboration could positively impact people's lives and businesses.

2.6.2 The Potential of XR Technology to Enhance Industry

XR technologies such as AR and VR have evolved rapidly in recent years and can potentially bring about revolutionary changes in the industry. The following is a brief overview of the possibilities of XR technology and the metaverse. Below are a few key points regarding the potential of XR technology and its relevance to the metaverse.

– ***Enhanced immersive experience:***

XR technology is an essential tool for providing immersive experiences. Virtual reality (VR) and augmented reality (AR) enable users to immerse themselves in realistic environments and have virtual experiences. Metaverse extends these immersive experiences and provides freedom of activity and interaction in virtual worlds.

– ***Collaboration in virtual space:***

Metaverse enables multiple users to collaborate and communicate in real-time in virtual space, and when combined with XR technology, can turn such virtual collaboration into a richer experience. For example, remote teams can use VR to hold meetings in virtual spaces, or AR can be used to share digital information in natural environments.

– ***Business activities in virtual space:***

Metaverse enables business activities to occur in virtual spaces. XR technology can build virtual stores and offices where products can be displayed and sold, and meetings and events can be held. This allows businesses to transcend geographical constraints and access new markets and customer segments.

– ***Fusion of Real and Virtual:***

Combining XR technology and the Metaverse fuses natural environments and virtual spaces. For example, users can use AR to receive virtual guidance and information while walking through an actual city or VR to try on natural products in a virtual space. This blurs the boundary between real and virtual, providing new experiences and services.

As described above, XR technology and the metaverse complement each other to provide richer experiences and business opportunities. It is expected to advance the convergence of physical and cyberspace and bring innovation to people's lives and businesses.

Virtual reality (VR) technology is one of the most exciting areas in the twenty-first century, combining innovative elements such as computer graphics and sensor technology to immerse the user in a virtual world. Its appeal lies in its ability to transcend the natural world's constraints, explore uncharted territory, and enjoy new experiences.

References

1. Autor, D., Salomons, A.: Is automation labor-displacing? Productivity growth, employment, and the labor share. Brook. Pap. Econ. Act. **2018**(1), 1–87 (2018)
2. Frey, C.B., Osborne, M.A.: The future of employment: how susceptible are jobs to computerisation? Technol. Forecast. Soc. Chang. **114**, 254–280 (2017)
3. Arntz, M., Gregory, T., Zierahn, U.: The Risk of Automation for Jobs in OECD Countries: A Comparative Analysis. OECD Social, Employment and Migration Working Papers, No. 189. OECD Publishing, Paris (2016)
4. Brynjolfsson, E., McAfee, A.: The Second Machine Age: Work, Progress, and Prosperity in a Time of Brilliant Technologies. WW Norton & Company (2014)
5. World Economic Forum: The Future of Jobs Report 2018. World Economic Forum, Geneva (2018)
6. Acemoglu, D., Restrepo, P.: Artificial Intelligence, Automation and Work. NBER Working Paper No. 24196 (2018)
7. Susskind, R., Susskind, D.: The Future of the Professions: How Technology Will Transform the Work of Human Experts. Oxford University Press (2015)
8. Bessen, J.: AI and Jobs: The Role of Demand. NBER Working Paper No. 24235 (2019)
9. McKinsey Global Institute: Jobs Lost, Jobs Gained: Workforce Transitions in a Time of Automation (2017)
10. Manyika, J., Lund, S., Chui, M., Bughin, J., Woetzel, J., Batra, P., Ko, R.: Jobs Lost, Jobs Gained: What the Future of Work Will Mean for Jobs, Skills, and Wages. McKinsey Global Institute (2017)
11. National Research Council: Successful K-12 STEM Education: Identifying Effective Approaches in Science, Technology, Engineering, and Mathematics. National Academies Press (2011)
12. OECD: PISA 2018 Results (Volume V): Effective Policies, Successful Schools. OECD Publishing (2019)
13. Darling-Hammond, L., Bransford, J. (eds.): Preparing Teachers for a Changing World: What Teachers Should Learn and Be Able to Do. Wiley (2005)

14. Becker, G.S.: Human Capital: A Theoretical and Empirical Analysis, with Special Reference to Education. University of Chicago Press (1993)
15. Fadel, C., Trilling, B.: 21st Century Skills: Learning for Life in Our Times. Wiley (2009)
16. Halverson, R., Shapiro, R.B.: Rethinking Education in the Age of Technology: The Digital Revolution and Schooling in America. Teachers College Press (2013)
17. Fullan, M.: The New Meaning of Educational Change. Teachers College Press (2016)
18. DiMaggio, P., Hargittai, E.: From the 'Digital Divide' to 'Digital Inequality': Studying Internet Use as Penetration Increases. Princeton University, Center for Arts and Cultural Policy Studies (2001)
19. Warschauer, M.: Technology and Social Inclusion: Rethinking the Digital Divide. MIT Press (2003)
20. van Dijk, J.A.: Digital divide research, achievements, and shortcomings. Poetics. **34**(4–5), 221–235 (2006)
21. Hargittai, E.: Second-level digital divide: differences in people's online skills. First Monday. **7**(4) (2002)
22. Norris, P.: Digital Divide: Civic Engagement, Information Poverty, and the Internet Worldwide. Cambridge University Press (2001)
23. Mossberger, K., Tolbert, C.J., Stansbury, M.: Virtual Inequality: Beyond the Digital Divide. Georgetown University Press (2003)
24. van Dijk, J.A.: The Deepening Divide: Inequality in the Information Society. Sage (2005)
25. Warschauer, M.: Reconceptualizing the digital divide. First Monday. **7**(7) (2002)
26. Van Deursen, A.J., Van Dijk, J.A.: The first-level digital divide shifts from inequalities in physical access to inequalities in material access. New Media Soc. **21**(2), 354–375 (2019)
27. Selwyn, N.: Reconsidering political and popular understandings of the digital divide. New Media Soc. **6**(3), 341–362 (2004)
28. Solove, D.J.: A taxonomy of privacy. Univ. Pa. Law Rev. **154**(3), 477–564 (2006)
29. Acquisti, A., Gross, R.: Imagined communities: awareness, information sharing, and privacy on the Facebook. In: Privacy Enhancing Technologies, pp. 36–58. Springer, Berlin/Heidelberg (2006)
30. Cavoukian, A., Jonas, J.: Privacy by Design: The Definitive Guide. Apress (2012)
31. Mayer-Schönberger, V.: Delete: The Virtue of Forgetting in the Digital Age. Princeton University Press (2009)
32. Nissenbaum, H.: Privacy in Context: Technology, Policy, and the Integrity of Social Life. Stanford University Press (2011)
33. Rotenberg, M., Scott, A.: Privacy in the Modern Age: The Search for Solutions. The New Press (2010)
34. Schneier, B.: Data and Goliath: The Hidden Battles to Collect Your Data and Control Your World. WW Norton & Company (2015)
35. Swire, P.P.: A model for when disclosure helps security: what is different about computer and network security? J. Telecommun. High Technol. Law. **10**, 209 (2012)
36. Oh, H., Rizo, C., Enkin, M., Jadad, A.: What is eHealth (3): a systematic review of published definitions. J. Med. Internet Res. **7**(1), e1 (2005)
37. Wootton, R.: Twenty years of telemedicine in chronic disease management-an evidence synthesis. J. Telemed. Telecare. **18**(4), 211–220 (2012)
38. Kvedar, J., Coye, M.J., Everett, W.: Connected health: a review of technologies and strategies to improve patient care with telemedicine and telehealth. Health Aff. **33**(2), 194–199 (2014)
39. Pagliari, C., Sloan, D., Gregor, P., Sullivan, F., Detmer, D., Kahan, J.P., Oortwijn, W.: What is eHealth (4): a scoping exercise to map the field. J. Med. Internet Res. **7**(1), e9 (2005)
40. Wu, R.C., Orr, M.S., Chignell, M., Straus, S.E.: Usability testing of a mobile health app: a case study of clinician input and patient-reported outcomes. JMIR mHealth uHealth. **1**(2), e10 (2013)

41. Agarwal, S., Perry, H.B., Long, L.A., Labrique, A.B.: Evidence on feasibility and effective use of mHealth strategies by frontline health workers in developing countries: systematic review. Trop. Med. Int. Health. **20**(8), 1003–1014 (2015)
42. Mosa, A.S., Yoo, I., Sheets, L.: A systematic review of healthcare applications for smartphones. BMC Med. Inf. Decision Making. **12**(1), 67 (2012)
43. Chib, A., Lin, S. (eds.): Mobile Health Solutions for Biobehavioral Research. Springer Science & Business Media (2013)
44. Istepanian, R.S.H., Jovanov, E., Zhang, Y.T. (eds.): m-Health: Fundamentals and Applications. Wiley (2016)
45. Schaltegger, S., Burritt, R.: Business cases and corporate engagement with sustainability: differentiating ethical motivations. J. Bus. Ethics. **147**(2), 241–259 (2018)
46. Hawken, P., Lovins, A., Lovins, L.H.: Natural Capitalism: The Next Industrial Revolution. Routledge (2013)
47. UN Environment Programme: Emissions Gap Report 2019. United Nations Environment Programme (2019)
48. McDonough, W., Braungart, M.: The Upcycle: Beyond Sustainability—Designing for Abundance. Macmillan (2013)
49. World Economic Forum: The Global Competitiveness Report 2018. World Economic Forum (2018)
50. GeSI (Global e-Sustainability Initiative): #SMARTer2030: ICT Solutions for 21st Century Challenges. Global e-Sustainability Initiative (2019)
51. Lovins, A.B., Lovins, L.H., Hawken, P.: A roadmap for natural capitalism. Harv. Bus. Rev. **77**(3), 145–158 (1999)
52. Cohen, B., Winn, M.I.: Market imperfections, opportunity and sustainable entrepreneurship. J. Bus. Ventur. **22**(1), 29–49 (2007)
53. Castells, M.: The Rise of the Network Society: The Information Age: Economy, Society, and Culture, vol. 1. Wiley (2011)
54. Turkle, S.: Alone Together: Why We Expect More from Technology and Less from Each Other. Basic Books (2011)
55. Wellman, B., Haythornthwaite, C.: The Internet in Everyday Life. Wiley (2002)
56. Rheingold, H.: The Virtual Community: Homesteading on the Electronic Frontier. MIT Press (2000)
57. Papacharissi, Z.: A Networked Self: Identity, Community, and Culture on Social Network Sites. Routledge (2011)
58. Giddens, A.: Modernity and Self-Identity: Self and Society in the Late Modern Age. Stanford University Press (1991)
59. Russell, S.J., Norvig, P.: Artificial Intelligence: A Modern Approach. Pearson (2016)
60. Goodfellow, I., Bengio, Y., Courville, A.: Deep Learning. MIT Press (2016)
61. Domingos, P.: The Master Algorithm: How the Quest for the Ultimate Learning Machine Will Remake Our World. Basic Books (2015)
62. Brynjolfsson, E., McAfee, A.: The Second Machine Age: Work, Progress, and Prosperity in a Time of Brilliant Technologies. WW Norton & Company (2017)
63. Bostrom, N.: Superintelligence: Paths, Dangers, Strategies. Oxford University Press (2014)
64. Chui, M., Manyika, J., Miremadi, M.: Where machines could replace humans—and where they can't (yet). McKinsey Q., 1–9 (2016)
65. Davenport, T.H., Ronanki, R.: Artificial intelligence for the real world. Harv. Bus. Rev. **96**(1), 108–116 (2018)
66. LeCun, Y., Bengio, Y., Hinton, G.: Deep learning. Nature. **521**(7553), 436–444 (2015)
67. Silver, D., Huang, A., Maddison, C.J., Guez, A., Sifre, L., Van Den Driessche, G., et al.: Mastering the game of Go with deep neural networks and tree search. Nature. **529**(7587), 484–489 (2016)
68. Atzori, L., Iera, A., Morabito, G.: The Internet of Things: a survey. Comput. Netw. **54**(15), 2787–2805 (2010)

69. Gubbi, J., Buyya, R., Marusic, S., Palaniswami, M.: Internet of Things (IoT): a vision, architectural elements, and future directions. Futur. Gener. Comput. Syst. **29**(7), 1645–1660 (2013)
70. Vermesan, O., Friess, P., Guillemin, P., Gusmeroli, S., Sundmaeker, H., Bassi, A., et al.: Internet of Things Strategic Research Roadmap. River Publishers (2011)
71. Lee, I., Lee, K.: The Internet of Things (IoT): applications, investments, and challenges for enterprises. Business Horizons. **58**(4), 431–440 (2015)
72. Haller, S., Karnouskos, S., Schroth, C.: The Internet of Things in an enterprise context. In: Internet of Things, pp. 14–28. Springer, Berlin/Heidelberg (2008)
73. Zanella, A., Bui, N., Castellani, A., Vangelista, L., Zorzi, M.: Internet of Things for smart cities. IEEE Internet Things J. **1**(1), 22–32 (2014)
74. Al-Fuqaha, A., Guizani, M., Mohammadi, M., Aledhari, M., Ayyash, M.: Internet of Things: a survey on enabling technologies, protocols, and applications. IEEE Commun. Surv. Tutorials. **17**(4), 2347–2376 (2015)
75. Ganti, R.K., Jayachandran, P.: Internet of Things (IoT): applications and challenges in technology and standardization. Wirel. Pers. Commun. **83**(3), 2349–2365 (2015)
76. Gluhak, A., Krco, S., Nati, M., Pfisterer, D., Mitton, N., Razafindralambo, T.: A survey on facilities for experimental Internet of Things research. IEEE Commun. Mag. **49**(11), 58–67 (2011)
77. Nakamoto, S.: Bitcoin: A Peer-to-Peer Electronic Cash System (2008). Retrieved from https://bitcoin.org/bitcoin.pdf
78. Tapscott, D., Tapscott, A.: Blockchain Revolution: How the Technology Behind Bitcoin Is Changing Money, Business, and the World. Penguin (2016)
79. Antonopoulos, A.M.: Mastering Bitcoin: Unlocking Digital Cryptocurrencies. O'Reilly Media, Inc (2014)
80. Swan, M.: Blockchain: Blueprint for a New Economy. O'Reilly Media, Inc (2015)
81. Casey, M.J., Vigna, P.: The Truth Machine: The Blockchain and the Future of Everything. St. Martin's Press (2018)
82. Mougayar, W.: The Business Blockchain: Promise, Practice, and Application of the Next Internet Technology. Wiley (2016)
83. Swan, M. (ed.): Blockchain: Blueprint for a New Economy. O'Reilly Media, Inc (2017)
84. Narayanan, A., Bonneau, J., Felten, E., Miller, A., Goldfeder, S.: Bitcoin and Cryptocurrency Technologies: A Comprehensive Introduction. Princeton University Press (2016)
85. Don Tapscott, A., Tapscott, A.: Blockchain Revolution: How the Technology Behind Bitcoin Is Changing Money, Business, and the World. Penguin (2017)
86. Sherman, W.R., Craig, A.B.: Understanding Virtual Reality: Interface, Application, and Design. Morgan Kaufmann (2019)
87. Burdea, G.C., Coiffet, P.: Virtual Reality Technology. Wiley (2003)
88. Milgram, P., Takemura, H., Utsumi, A., Kishino, F.: Augmented reality: a class of displays on the reality-virtuality continuum. In: Telemanipulator and Telepresence Technologies, vol. 2351, pp. 282–292. International Society for Optics and Photonics (1995)
89. Azuma, R.T.: A survey of augmented reality. Presence Teleoper. Virtual Environ. **6**(4), 355–385 (1997)
90. Billinghurst, M., Kato, H.: Collaborative augmented reality. Commun. ACM. **45**(7), 64–70 (2002)
91. Milgram, P., Kishino, F.: Taxonomy of mixed reality visual displays. IEICE Trans. Inf. Syst. **77**(12), 1321–1329 (1994)
92. Cruz-Neira, C., Sandin, D.J., DeFanti, T.A.: Surround-screen projection-based virtual reality: the design and implementation of the CAVE. In: Proceedings of the 20th Annual Conference on Computer Graphics and Interactive Techniques, pp. 135–142 (1993)
93. Azuma, R., Baillot, Y., Behringer, R., Feiner, S., Julier, S., MacIntyre, B.: Recent advances in augmented reality. IEEE Comput. Graph. Appl. **21**(6), 34–47 (2001)
94. Rizzo, A.A., Kim, G.J.: A SWOT analysis of the field of virtual reality rehabilitation and therapy. Presence Teleoper. Virtual Environ. **14**(2), 119–146 (2005)

95. Nielsen, M.A., Chuang, I.L.: Quantum Computation and Quantum Information. Cambridge University Press (2010)
96. Ladd, T.D., Jelezko, F., Laflamme, R., Nakamura, Y., Monroe, C., O'Brien, J.L.: Quantum computers. Nature. **464**(7285), 45–53 (2010)
97. Preskill, J.: Quantum computing in the NISQ era and beyond. Quantum. **2**, 79 (2018)
98. Kitaev, A.Y., Shen, A.H., Vyalyi, M.N.: Classical and Quantum Computation, vol. 47. American Mathematical Society (2002)
99. Montanaro, A.: Quantum algorithms: an overview. npj Quantum Inf. **2**(1), 1–10 (2016)
100. Wilde, M.M.: Quantum Information Theory. Cambridge University Press (2017)
101. Harrow, A.W., Montanaro, A.: Quantum computational supremacy. Nature. **549**(7671), 203–209 (2017)
102. Rieffel, E.G., Polak, W.H.: Quantum Computing: A Gentle Introduction. MIT Press (2011)
103. Yanofsky, N.S., Mannucci, M.A.: Quantum Computing for Computer Scientists. Cambridge University Press (2008)
104. Coles, P.J., Berta, M., Tomamichel, M., Wehner, S.: Entropic uncertainty relations and their applications. Rev. Mod. Phys. **89**(1), 015002 (2018)

Chapter 3
Introduction to Metaverse with a Focus on Second Life (SL) and Japan's Island There

Dana M. Barry and Hideyuki Kanematsu

Abstract This chapter introduces Metaverse and then focuses on SL and Japan's island located there. Metaverse provides a fully immersive and interconnected virtual world where individuals can work, live, etc. and interact in a digital environment that closely resembles the real world. On the other hand, SL is a three-dimensional world where avatars perform certain tasks on behalf of us. Details about Metaverse and Second Life are provided. Also, Japan's island in SL is described in terms of the virtual classrooms and teaching/learning opportunities that it provides. The country's island is owned by Nagaoka University of Technology and researchers there built virtual buildings containing virtual classrooms for various projects and activities (which will be discussed in Chap. 4). The classrooms include red chairs, tables, a podium, whiteboards, a recording box to record chat dialogues, and microphones for those using the speech function to communicate. This arrangement gives the participants a sense of reality because the virtual classrooms resemble those seen in real life. Students use avatars in the virtual environment to carry out activities on behalf of them. Their avatars construct items such as houses by preparing prims (three dimensional objects like cubes) in a space referred to as a sandbox, located in front of the virtual classroom.

Keywords Metaverse · Second Life · Avatars · Virtual classrooms

D. M. Barry (✉)
Clarkson University, Potsdam, NY, USA
e-mail: dbarry@clarkson.edu

H. Kanematsu
National Institute of Technology, Suzuka College, Suzuka, Mie, Japan
e-mail: hideyuki.kanematsu@bioenglab.org

D. M. Barry, H. Kanematsu (eds.), *Applications of Metaverse and Virtual Reality to Creative Education and Industry*, Intelligent Systems Reference Library 267, https://doi.org/10.1007/978-981-96-3341-8_3

3.1 Metaverse

Cyberspace has been changing throughout the 1990s and 2000s, especially when the Internet was widely in use. As a result of these changes, Metaverse was formed. It has various definitions. However, for our purpose, it is a large, sustainable, and interconnected cyberworld that simulates our existing one. The name Metaverse is the combination of two words: "meta" (which means beyond or to climb) and "universe." It was first used in Neil Stevenson's science fiction novel *Snow Crash* in 1992 [1]. The novel includes a world with virtual and reality interactions through a variety of social activities. In this world, individuals are represented by avatars that carry out tasks on their behalf. Gradually with time, the role of an avatar has changed to reflect the user's ego [2, 3]. This avatar wears luxury clothes and has a suitable position for a job in Metaverse. The younger generation may think that their identity in virtual space and reality are the same.

Metaverse relies on extended reality (XR). In terms of technology, XR is related to VR (virtual reality), AR (augmented reality), and MR (mixed reality) [4–7]. Virtual reality provides experiences, so the users feel as if they are in special places without physical limitations. It allows a new reality based on 360-degree images. It covers the entire field of view, provides an immersive feeling, and is good for long term content. AR superimposes virtual objects on real ones. For example, it overlays computer generated images, sounds, etc. into a real-world environment. Augmented reality (AR) has relatively simple hardware like glasses and is suitable for short content. MR is a mixed reality technology that integrates VR and AR. In addition, 3-D modeling and animation technology are used to create detailed and realistic virtual objects and environments. Metaverse also involves Block Chain Technology, which provides a massive layer of security.

The Metaverse implementation process contains a design phase, a model-training phase, an operation phase, and an evaluation phase [8–14]. The design phase includes goals, ideas, and concepts for the design, development time, cost, and risk estimates, etc. In the model-training phase, data analysis and science methods are used. The operation phase considers simulations, job scheduling, network environments, etc. The evaluation phase deals with the content and the overall Metaverse implementation process.

COMPONENT: HARDWARE

The following components of Metaverse are presented and briefly described: hardware, content, challenges, applications, and trends [15–47]. The important hardware of Metaverse is the Head-Mounted Display (HMD). It blocks the view to enable immersive participation. Also, it shows an image through the display and plays sound through a speaker. Several issues with the HMD are that it is bulky, expensive, and has a short battery life. Other hardware items include a Hand-Based Input Device, a Non-Hand Based Input Device, and Motion Input Device. A Hand Based Input Device gives individuals the texture of real objects. Using real props in a virtual environment enhances a user's experience. This device can be attached to the hand. Non-Hand Based Input Devices include those for eye tracking, head tracking, voice input, etc. For example, eye tracking predicts eye movement when the users move their eyes without turning their heads. Motion Input Devices involve both passive and active methods. The passive method delivers a sense to the user with a fixed scenario. However, the active method provides different forms of realism like walking a 360-degree rotation.

COMPONENT: CONTENT

Content is the important component that maintains the Metaverse. It is used to provide an immersive experience through organized stories and events created by the user. Metaverse is an active medium. It involves creating interactive and immersive stories that engage participants. It simulates a realm where creativity has no bounds, limitless interactions, and where individuals have the power to shape their identity. Metaverse contains interconnected platforms and technologies. It provides many possibilities for us to communicate, collaborate, work, learn, have fun, etc. An example of living in a story in Metaverse, is the game Fortnite. It hosts live events featuring celebrities, musicians, etc. Participants join these events with their avatars for dancing and having fun. Content is created by using the paradigm shift method and a method to reuse existing content. A paradigm shift is a change in the perceptions and beliefs of a person or a group of people. An example is requiring people to wear a facemask in public during the pandemic. It should also be mentioned that some content areas that require environmental design are scenes, color and lighting, audio, etc.

COMPONENT: CHALLENGES

Metaverse includes challenges and problems. Users can suffer from motion sickness like cyber motion sickness due to an imbalance of visual information obtained from human eyes and organs. There are issues like physical fatigue, headset weight, and movement injuries. Areas of concern relate to ethics, data security, safety, exposure to undesirable behaviors like harassment of users, unregulated gambling, bullying, and threats of violence, etc. It should be mentioned that excessive immersion leads to psychological problems like separation from reality. Also, negative feelings and emotions that occur in the Metaverse can extend to the real world, which can lead to social problems.

COMPONENT: APPLICATIONS

Metaverse is a large, shared virtual space, where users can interact and engage with virtual objects, virtual environments, and with each other. It served an important role during the COVID pandemic to reduce the impact of social isolation. Metaverse has a variety of applications. Through simulations, it is used for educational purposes, museum tours, and to depict real world tasks and situations. A few applications are provided. Metaverse can be used to create virtual field trips that allow students to explore and learn about history and special places /events. This could be useful for students who are not able, or have no opportunity, to go on physical field trips due to a lack of money. Games are a very popular platform in Metaverse. This virtual world can also be used for entertainment, social networking, business, etc. As for entertainment, the Metaverse allows users to attend concerts, conferences, and other exciting events taking place around the world. It may provide a more rewarding experience than one in real life. Regarding healthcare, Metaverse could be used to create virtual clinics and hospitals where patients can access healthcare services in a convenient way, especially those who do not have easy access to physical healthcare facilities. For real estate, Metaverse has the potential to offer virtual spaces in which users can explore and interact with digital representations of real-world properties in 3-D form from anywhere in the world. The Metaverse may also be used to build virtual factories and assembly lines to test and evaluate creative manufacturing methods and technologies. Also, virtual prototypes and simulations of products may be built there. The Metaverse has the potential to change the way transportation systems are designed, planned, and evaluated. This can be done by using virtual environments to simulate and test different scenarios for autonomous vehicles. This allows engineers to evaluate the performance of different control systems, etc. For sports, the Metaverse offers a virtual world where people may participate in and watch sporting events. It might even provide a new system for marketing and item sales. In addition, military training in Metaverse helps by creating simulations with a variety of situations for shooting, strategic action, and observational training.

COMPONENT: TRENDS
Starting in 2024, Metaverse emerges as a digital frontier that interwinds the virtual and physical realms. Artificial intelligence (AI) appears to be a catalyst for transforming experiences within the Metaverse. It promotes heightened realism and immersion. High—fidelity avatars (those with realistic looks and qualities) and dynamic scene digitization will be created. Metaverse will go beyond a virtual entertainment center and influence diverse sectors. It will explore trends in virtual economies, urban planning, AI integration, and gamification. It plans to redefine digital experiences. As for virtual economies, there is a shift in the marketplace where digital assets gain tangible value that impacts creators, investors, and consumers. Regarding architecture and urban planning, it serves as a testing ground for urban planning. It uses data-driven decision making (often used with machine learning methods) and provides a virtual blueprint for real-world architecture and planning. The transition from 2D to 3D interactions changes the way we connect online. Gaming economies thrive in Metaverse with platforms like Fortnite. Metaverse also becomes a space for research, training, and experimentation. As the Metaverse further integrates into our daily lives, policy considerations become very important to protect participants' safety, security, and privacy.

3.2 Second Life

Second Life (SL) is an online multimedia platform that allows people to create an avatar for themselves and to interact with other users and user-created content within a multi-user virtual world. It was launched in June 2023 because of Philip

Rosedale's founding of Linden Lab in 1999, with the goal of creating hardware that would immerse people in a virtual world. SL provides a unique online environment where users can express themselves, create, and interact, without being constrained by traditional gaming objectives. Highlights of SL include the creation of avatars to carry out tasks on our behalf, social interactions where users can meet other users to socialize and participate in individual and group activities, and the building and creating of various items. This platform mainly features 3D-based user-generated content. Second Life also has its own virtual currency called the Linden Dollar, which is exchangeable with real-world currency. Studies related to SL are provided [48–76].

VARIOUS STUDIES RELATED TO SECOND LIFE

Cross Cultural Exchange: Second Life has a global reach and provides for cross-cultural exchange by using multiple modes of communication in real and virtual worlds. Therefore, it provides an opportunity to examine cross-cultural engagement. This project used the Cultural Historical Activity Theory. This theory was developed by cultural historians in the early twentieth century as they were exploring how different cultures relate to each other. It also involves Heyward's Model of Intercultural Literacy. Intercultural literacy is the understandings, attitudes, competencies, and identities that enable effective participation in a cross-cultural setting. The theory and model were used to analyze findings from an exploratory study examining the construction of cultural identity and the development of intercultural literacy for 29 participants in Second Life. The authors found that the participants engaged in many activities like language classes, etc. The results showed that participation in SL enhanced participants intercultural literacy in many ways. Their SL experience fostered the use of multiple languages, cross-cultural encounters and friendships, greater awareness of insider perspectives, and openness towards new viewpoints. Also, an avatar's appearance represented the participant's cultural identity.

Psychological Needs: A study was carried out to find out SL users' psychological needs and motivations for the use of the virtual world. This information could be important for designers and researchers working on 3D virtual worlds. SL was selected as the virtual environment for the study because many different activities are available, and the users can develop new activities and personalize the experience. Data was collected by using a global online questionnaire that targeted experienced SL users of about 40 years of age. Responses received totaled 258. The results showed clear patterns of how psychological needs of the participants are generally satisfied in the real world and the virtual world. Quantitative results suggested the feelings of autonomy, self-esteem, physical thriving, pleasure, stimulation, and relatedness were strong needs. The authors discovered that autonomy, physical thriving, and money luxury needs were met better in the virtual world, while competence, relatedness, security, and popularity influence were better met in the users' daily lives. Qualitative data highlighted relatedness needs as motivations for the use of SL and revealed 5 control themes for it.: SL as self-therapy, a source of instant pleasure, a liberation from social norms, a tool for self-expression, and as exploration and novelty.

Medical and Health Education: Second Life has a good number of medical and health education projects. For example, Ohio University has a nutrition game in SL. Visitors learn about the impact fast food has on health. Players experiment with different eating styles in simulated fast-food restaurants. They learn about the long- and short-term health impacts of their choices. Individuals making healthy choices get a high score for the game.

Social Work Education: A study was conducted to determine the effectiveness of Social Work education in SL. It evaluated student perspectives of the educational value of learning experiences in SL while enrolled in an undergraduate introduction to Social Welfare and Social Work class. Analyses of surveys and journal content showed that the students found the virtual world learning to be useful in teaching social work values, skills, and related knowledge. It also provided an emotional and thought-provoking experience.

Hospitality, Tourism, and More: Second Life allows visitors to tour virtual hotels and virtual islands. Some travel companies are set up in SL, so individuals can book real-life holidays. Also, conferences and event companies have virtual and exhibition centers in SL and promote exciting locations for field trips to museums, foreign countries, etc. Researchers Kaplan and Haenlein found different ways in which companies can make use of SL. They include advertising, communications, virtual product sales, marketing research, human resource management, and internal process management.

3.3 Japan's Island in Second Life

Japan's island in Second Life is owned by Nagaoka University of Technology (NUT), which is in Nagaoka, Niigata, Japan [77]. This is a public university that was founded in 1976. It is one of two universities of technology in Japan. The other one is Toyohashi University of Technology in Aichi. NUT offers undergraduate and graduate programs in areas including engineering, science, and technology. Also, it collaborates closely with industry partners, government agencies, and finance institutions. Japan's island in SL is described in terms of the virtual classrooms and teaching/learning opportunities that it provides. The country's island is owned by Nagaoka University of Technology and researchers there built virtual buildings containing virtual classrooms for various projects and activities (which will be discussed in Chap. 4). The classrooms include red chairs, tables, a podium, whiteboards, a recording box to record chat dialogues, and microphones for those using the speech function to communicate. This arrangement gives the participants a sense of reality because the virtual classrooms resemble those seen in real life. Students use avatars in the virtual environment to carry out activities on behalf of them. Their avatars construct items such as houses by preparing prims (three dimensional objects like cubes) in a space referred to as a sandbox, located in front of the virtual

Fig. 3.1 An instructor's avatar gives a lecture to students in a virtual classroom on Japan's island in SL

Fig. 3.2 Instructors observe students as they make primitives (3-D objects) for upcoming building projects in the sandbox area in SL

classroom. NOTE: Students in Second Life can work from anywhere in the world, at any time, and at their own pace. Two figures are provided. Figure 3.1 is a display of the virtual classroom on Japan's island in Second Life [78]. The instructor's avatar is standing at the podium giving a lecture by using the speech function. A microphone is visible in the photo. The students' avatars are attentively seated in red chairs. Figure 3.2 shows the sandbox area (outside of the virtual classroom) where building projects take place [79]. Instructors observe students as they make primitives (three-dimensional objects like cubes of various shapes, sizes, color, texture, etc.) for an upcoming building project.

3.4 Conclusions

This chapter introduces Metaverse and then focuses on SL and Japan's island located there. Metaverse provides a fully immersive and interconnected virtual world where individuals can work, live, etc. and interact in a digital environment that closely resembles the real world. On the other hand, SL is a three-dimensional world where avatars perform certain tasks on behalf of us. Details about Metaverse and Second Life are provided. Also, Japan's island in SL is described in terms of the virtual classrooms and teaching/learning opportunities that it provides. The country's island is owned by Nagaoka University of Technology and researchers there built virtual buildings containing virtual classrooms for various projects and activities (which will be discussed in Chap. 4). The classrooms include red chairs, tables, a podium, whiteboards, a recording box to record chat dialogues, and microphones for those using the speech function to communicate. This arrangement gives the participants a sense of reality because the virtual classrooms resemble those seen in real life. Students use avatars in the virtual environment to carry out activities on behalf of them. Their avatars construct items such as houses by preparing prims (three dimensional objects like cubes) in a space referred to as a sandbox, located in front of the virtual classroom. It should be mentioned that in Second Life, students can work from anywhere, at any time, and at their own pace.

References

1. Gao, H., Chong, A.Y.L., Bao, H.: Metaverse: literature review, synthesis, and future research agenda. J. Comput. Inf. Syst., 1–21 (2023). https://doi.org/10.1080/08874417.2023.2233455
2. Park, S.-M., Kim, Y.-G.: A metaverse: taxonomy, components, applications, and open challenges. IEEE Access. **10**, 4209–4251 (2022). https://doi.org/10.1109/ACCESS.2021.3140175
3. Cagnina, M.R., Poian, M.: How to Compete in the Metaverse: The Business Models in Second Life. U. of Udine Economics working Paper No. 01-2007, p. 4 (2007)
4. Duan, H., Li, J., Fan, S., Lin, Z., Wu, X., Cai, W.: Metaverse for social good: a university campus prototype. Proc. 29th ACM Int. Conf. Multimedia, 153–161 (2021)
5. Choi, H.-S., Kim, S.-H.: A content service deployment plan for metaverse museum exhibitions: centering on the combination of beacons and HMDs. Int. J. Inf. Manag. **37**(1), 1519–1527 (2017)
6. Suzuki, S.-N., Kanematsu, H., Barry, D.M., Ogawa, N., Yajima, K., Nakahira, K.T., et al.: Virtual experiments in metaverse and their applications to collaborative projects: the framework and its significance. Proc. Comput. Sci. **176**, 2125–2132 (2020)
7. Ryskeldiev, B., Ochiai, Y., Cohen, M., Herder, J.: Distributed metaverse: Creating decentralized blockchain-based model for peer-to-peer sharing of virtual spaces for mixed reality applications. Proc. 9th Augmented Hum. Int. Conf., 1–3 (2018)
8. Jaynes, C. Sealers, W. B. K., Calvert, K., Fei, Z. and J. Griffioen (2003) The metaverse: a networked collection of inexpensive self-configuring immersive environments. Proc. Workshop Virtual Environ. (EGVE), 115-124.
9. Ondrejka, C.: Escaping the gilded cage: user created content and building the metaverse. NYL Sch. L. Rev. **49**, 81 (2004)

10. Collins, C.: Looking to the future: higher education in the metaverse. Educ. Rev. **43**(5), 51–63 (2008)
11. Wright, M., Ekeus, H., Coyne, R., Stewart, J., Travlou, P., Williams, R.: Augmented duality: overlapping a metaverse with the real world. Proc. Int. Conf. Adv. Comput. Entertainment Technol. (ACE), 263–266 (2008)
12. Schlemmer, E., Trein, T.D., Cristoffer, O.: The metaverse: telepresence in 3D avatar-driven digital-virtual worlds Tic. Revista d'innovaci? Educativa. **2**, 26–32 (2009)
13. Messinger, P.R., Stroulia, E., Lyons, K., Bone, M., Niu, R.H., Smirnov, K., et al.: Virtual worlds: past present and future: new directions in social computing. Decis. Support. Syst. **47**(3), 204–228 (2009)
14. Hazan, S.: Musing the metaverse. In: Heritage in the Digital Era. Multi-Science Publishing, Brentwood (2010)
15. Papagiannidis, S., Bourlakis, M.: Staging the new retail drama: at a metaverse near you! J. Virtual Worlds Res. **2**(5), 425–446 (2010)
16. Forte, M., Lercari, N., Galeazzi, F., Borra, D., D.: Metaverse communities and archaeology: the case of Teramo. Proc. EuroMed, 79–84 (2010)
17. Cunningham, T.C.: Marching Toward the Metaverse; Strategic Communication Through the New Media, a Monograph by T.C. Cunningham (2010)
18. Owens, D., Mitchell, A., Khazanchi, D., Zigurs, I.: An empirical investigation of virtual world projects and metaverse technology capabilities. ACM SIGMIS Database Database Adv Inf. Syst. **42**(1), 74–101 (2011)
19. Tonéis, C.N.: Puzzles as a creative form of play in metaverse. J. Virtual Worlds Res. **4**(1) (2011)
20. Abu-Salih, B.: Meta ontology: toward developing an ontology for the metaverse. Front. Big Data. **5**, 998648 (2022)
21. Alaimo, C., Kallinikos, J., Valderrama, E.: Platforms as service ecosystems: lessons from social media. J. Inf. Technol. **35**(1), 25–48 (2020)
22. Augenstein, D., Morschheuser, B.: Understanding human factors in the metaverse—an autonomous driving experiment. In: Proceeding of the European Conference on Information Systems, Research in Progress, Timisoara, Romania (2022)
23. Avital, M.: Peer review: toward a blockchain-enabled market-based ecosystem. Commun. Assoc. Inf. Syst. **42**(1), 646–653 (2018)
24. Bajwa, D.S., Lewis, L.F., Pervan, G., et al.: The adoption and use of collaboration information technologies: international comparisons. J. Inf. Technol. **20**(2), 130–140 (2005)
25. Bao, X., Shou, M., Yu, J., et al.: Exploring metaverse: affordances and risks for potential users. In: Proceeding of the International Conference on Information Systems, Copenhagen, Denmark (2022)
26. Berente, N., Hansen, S., Pike, J.C., et al.: Arguing the value of virtual worlds: patterns of discursive sensemaking of an innovative technology. MIS Q. **35**(3), 685–709 (2011) Scopus
27. Chaturvedi, A., Dolk, D., Drnevich, P.: Design principles for virtual worlds. MIS Q. **35**(3), 673–684 (2011)
28. Chen, T., Zhou, H., Yang, H., et al.: A review of research on metaverse defining taxonomy and adaptive architecture. In: 5th International Conference on Pattern Recognition and Artificial Intelligence (PRAI), Chengdu, China, 2022, pp. 960–965 (2022)
29. Cheng, X., Zhang, S., Liu, W., et al.: Understanding visitors' metaverse and in-person tour intentions during the COVID-19 pandemic: a coping perspective. In: Proceeding of the Hawaii International Conferences System Sciences, Maui, HI, 2023, pp. 554–562 (2023)
30. Davis, A., Murphy, J., Owens, D., et al.: Avatars, people, and virtual worlds: foundations for research in metaverses. J. Assoc. Inf. Syst. **10**(2), 90–117 (2009)
31. Dincelli, E., Yayla, A.: Immersive virtual reality in the age of the metaverse: a hybrid-narrative review based on the technology affordance perspective. J. Strateg. Inf. Syst. **31**(2), 101717 (2022) Scopus
32. Dionisio, J.D.N., Iii, W.G.B., Gilbert, R.: 3D virtual worlds and the metaverse: current status and future possibilities. ACM Comput. Surv. **45**(3), 34–38 (2013)

33. Drummer, D., Neumann, D.: Is code law? Current legal and technical adoption issues and remedies for blockchain-enabled smart contracts. J. Inf. Technol. **35**(4), 337–360 (2020)
34. Kim, J.: Advertising in the metaverse: research Agenda. J. Interact. Advert. **21**(3), 141–144 (2021)
35. Lee, H., Woo, D., Yu, S.: Virtual reality metaverse system supplementing remote education methods: based on aircraft maintenance simulation. Appl. Sci. **12**(5), 2667 (2022)
36. Wang, Y., et al.: A survey on metaverse: fundamentals, security, and privacy. IEEE Commun. Surv. Tutorials. **25**(1), 319–352 (2022)
37. Samala, A.D., et al.: Metaverse technologies in education: a systematic literature review using PRISMA. Int. J. Emerg. Technol. Learn. **18**(05), 231–252 (2023)
38. Hatzilygeroudis, I.: Metaverse. Encyclopedia. **2**(1), 486–497 (2022)
39. Bale, A.S., et al.: A comprehensive study on metaverse and its impacts on humans. Adv. Hum. Comput. Interact. **2022**, 1–11 (2022)
40. Lukava, T., Morgado Ramirez, D.Z., Barbareschi, G.: Two sides of the same coin: accessibility practices and neurodivergent users' experience of extended reality. J. Enabling Technol. **16**(2), 75–90 (2022)
41. Taylor, C.R.: Research on advertising in the metaverse: a call to action. Int. J. Advert. **41**(3), 383–384 (2022)
42. Dionisio, J.D.N., Burns III, W.G., Gilbert, R.: 3D virtual worlds and the metaverse: current status and future possibilities. ACM Comput. Surv. **45**, 1–38 (2013)
43. Gonzalez, M.A., et al.: Virtual worlds. Opportunities and challenges in the 21st century. Proc. Comput. Sci. **25**, 330–337 (2013)
44. Peukert, C., et al.: Metaverse: how to approach its challenges from a BISE perspective. Bus. Inf. Syst. Eng. **64**(4), 401–406 (2022)
45. Periyasami, S., Periyasamy, A.P.: Metaverse as a future promising platform business model: case study on fashion value chain. Business. **2**(4), 527–545 (2022)
46. Hollensen, S., Kotler, P., Opresnik, M.O.: Metaverse–the new marketing universe. J. Bus. Strateg. **44**(3), 119–125 (2022)
47. Henz, P.: The psychological impact of the Metaverse. Discov. Psychol. **2**(1), 47 (2022)
48. Chow, A., Andrews, S., Trueman, R.: A "Second Life": can this online, virtual reality world be used to increase the overall quality of learning and instruction in graduate distance learning programs? In: Simonson, M., et al. (eds.) Proceedings of Association for Educational Communications and Technology 2007, Volume 2, pp. 86–94. AECT, Anaheim (2007)
49. De Lucia, A., Francese, R., Passero, I., Tortora, G.: Development and evaluation of a virtual campus on Second Life: the case of second DMI. Comput. Educ. **52**(2009), 220–233 (2009)
50. deWinter, J., Vie, S.: Press enter to "say": using Second Life to teach critical media literacy. Comput. Compos. **25**(3), 313–322 (2008)
51. Brennen, B., dela Cerna, E.: Journalism in second life. Journal. Stud. **11**(4), 546–554 (2010)
52. Partala, T.: Psychological needs and virtual worlds: case Second Life. Int. J. Hum.-Comput. Stud. **69**(12), 787–800 (2011)
53. Greenberg, J., Nepkie, J., Pence, H.E.: The SUNY Oneonta Second Life music project. J. Educ. Technol. Syst. **37**(3), 251–258 (2009)
54. Diehl, W.C., Prins, E.: Unintended outcomes in *Second Life*: intercultural literacy and cultural identity in a virtual world. Lang. Intercult. Commun. **8**(2), 101–118 (2008)
55. Jennings, N., Collins, C.: Virtual or virtually U: educational institutions in Second Life. Int. J. Soc. Sci. **2**(3), 180–186 (2007)
56. Kluge, S., Riley, L.: Teaching in virtual worlds: opportunities and challenges. J. Issues Inf. Sci. Inf. Technol. **5**, 127–135 (2008)
57. Baker, S.C., Wentz, R.K., Woods, M.M.: Using virtual worlds in education: Second Life® as an educational tool. Teach. Psychol. **36**(1), 59–64 (2009)
58. Bainbridge, W.S.: The scientific research potential of virtual worlds. Science. **317**, 472–476 (2007)

59. Bente, G., Rüggenberg, S., Krämer, N.C., Eschenburg, F.: Avatar-mediated networking: increasing social presence and interpersonal trust in net-based collaborations. Hum. Commun. Res. **34**, 287–318 (2008)
60. Blascovitch, J.: Social influence within immersive virtual environments. In: Schroeder, R. (ed.) The Social Life of Avatars: Presence and Interaction in Shared Virtual Environments, pp. 127–145. Springer, London (2002)
61. Abdallah, S., Douglas, J.: Students' first impression of Second Life: a case from the United Arab Emirates. Turk. Online J. Dist. Educ. **11**(3), 183–192 (2010)
62. Andreas, K., Tsiatsos, T., Terzidou, T., Pomportsis, A.: Fostering collaborative learning in Second Life: metaphors and affordances. Comput. Educ. **55**(2), 603–615 (2010)
63. Blasing, M.T.: Second language in Second Life: exploring interaction, identity and pedagogical practice in a virtual world. Slavic East Eur. J. **54**(1), 96–117 (2009)
64. Broadribb, S., Carter, C.: Using Second Life in human resource development. Br. J. Educ. Technol. **40**(3), 547–550 (2009)
65. Burgess, M.L., Slate, J.R., Roja-LeBouef, A., LaParairie, K.: Teaching and learning in Second Life: using the Community of Inquiry (CoI) model to support online instruction with graduate students in instructional technology. Internet High. Educ. **13**(1–2), 84–88 (2010)
66. Cargill-Kipar, N.: My dragonfly fly's upside down! Using Second Life in multimedia design to teach students programming. Br. J. Educ. Technol. **40**(3), 539–542 (2009)
67. Cheong, D.: The effects of practice teaching sessions in second life on the change in pre-service teachers' teaching efficacy. Comput. Educ. **55**(2), 868–880 (2010)
68. Good, J., Howland, K., Thackray, L.: Problem-based learning spanning real and virtual words: a case study in Second Life. ALT-J. **16**(3), 163–172 (2008)
69. Guadagno, R.E., Muscanell, N.L., Okdie, B.M., Burk, N.M., Ward, T.B.: Even in virtual environment women shop and men build: a social role perspective on Second Life. Comput. Hum. Behav. **27**(1), 304–308 (2011)
70. Haycock, K., Kemp, J.W.: Immersive learning environments in parallel universes: learning through Second Life. Sch. Libr. Worldw. **14**(2), 89–97 (2008)
71. Ho, C.M.L., Ong, A.M.H.: Towards evaluative meaning-making through enactive role play: the case of pre-tertiary students in Second Life. J. Appl. Linguist. **4**(2), 171–194 (2007)
72. Hornik, S., Thornburg, S.: Really engaging accounting: Second Life as a learning platform. Issues Account. Educ. **25**(3), 361–378 (2010)
73. Jarmon, L., Traphagan, T., Mayrath, M.: Understanding project-based learning in Second Life with a pedagogy, training, and assessment trio. Educ. Media Int. **45**(3), 175–176 (2008)
74. Mahon, J., Bryant, B., Brown, B., Kim, M.: Using Second Life to enhance classroom management practice in teacher education. Educ. Media Int. **47**(2), 121–134 (2010)
75. Nie, M., Roush, P., Wheeler, M.: Second Life for digital photography: an exploratory study. Contemp. Educ. Technol. **1**(3), 267–280 (2010)
76. Wang, Y., Braman, J.: Extending the classroom through Second Life. J. Inf. Syst. Educ. **20**(2), 235–247 (2009)
77. Nagaoka University of Technology—Wikipedia
78. Barry, D.M., et al.: Virtual workshop for creative teaching of STEM courses. Proc. Comput. Sci. **126**, 927–936 (2018)
79. Barry, D.M., et al.: Problem-based learning in Second Life. Int. J. Modern Educ. Forum. **3**(1) (2014)

Chapter 4
Creative Problem-Based Learning (PBL) Activities Successfully Carried Out in Second Life (SL)

Dana M. Barry and Hideyuki Kanematsu

Abstract This chapter describes the teaching approach called problem-based learning (PBL) and discusses the creative ways it was used to carry out successful activities in virtual classrooms on Japan's island in Second Life (SL). Problem-based learning provides students with challenging, ill-structured problems related to their daily lives. The students receive guidance from their teachers and work cooperatively in a group where they hold brainstorming sessions to seek solutions to the problems. The authors are researchers who successfully carried out PBL activities with students in SL. This chapter presents and describes some of their activities in detail. For one example, students are asked to design and build the house of the future during the global warming era. In another one, students are asked to design and build the eco car of the future. An eco car is energy efficient, safe, and environmentally friendly. In a third activity about nuclear safety, students are asked to find out what kind of metals can effectively protect human beings from radiation. For an activity about renewable energy, the students decided to build a special airplane for wind energy. The final activity is about a virtual workshop for elementary school teachers (teachers of young children). The workshop provided them with creative teaching techniques to engage and turn students on to learning. The teachers were asked to select one of the creative teaching methods and use it to prepare a creative activity for their students and to share it with the SL workshop participants.

Keywords Problem-based learning (PBL) · Second Life (SL) · Virtual classrooms · PBL projects in SL · Creative education

D. M. Barry (✉)
Clarkson University, Potsdam, NY, USA
e-mail: dbarry@clarkson.edu

H. Kanematsu
National Institute of Technology, Suzuka College, Suzuka, Mie, Japan
e-mail: hideyuki.kanematsu@bioenglab.org

D. M. Barry, H. Kanematsu (eds.), *Applications of Metaverse and Virtual Reality to Creative Education and Industry*, Intelligent Systems Reference Library 267, https://doi.org/10.1007/978-981-96-3341-8_4

4.1 Problem-Based Learning and Studies About It

Problem-based learning (PBL) is a student-centered instructional approach. It was originally developed in the mid-1960s as a useful instructional alternative for conventional teaching in medical education. This method does not focus on solving problems with definite solutions. Instead, it allows students to experience enhanced communication and collaboration. Using PBL, students become active learners because they are asked to solve real-world problems. The students work in small collaborative groups and learn what they need to know to solve a problem. The teacher acts as a facilitator to guide the students in developing the cognitive skills required for problem solving and collaboration. Many studies have confirmed that PBL improves the performance of students in medical education and in various other areas including science, engineering, business studies, education, etc. [1–22]. Wang and Zhao carried out a study to find out the learning experiences of 18 pre-service teachers and how the instructor was affected when implementing PBL into a course called Principles of Instruction [1]. Methods for collecting data included reflective reports, interviews, observations of participants, students' reports, a questionnaire, pre- and post-tests, etc. The findings indicated that pre-service teachers' professional knowledge, learning engagement, reflective abilities, and teamwork were enhanced through the PBL approach. Also, the instructors' quality of teaching improved. In addition, Yew and Goh reviewed articles that suggested PBL is an effective teaching and learning approach, especially when it is evaluated for long-term retention and applications [23].

A study was also carried out to find the effective learning behavior in problem-based learning [23–42]. The authors (Ghani, Rahim, Yusoff, & Hadie) mention that problem-based learning emphasizes learning behavior that leads to critical thinking, problem-solving, communication, and collaborative skills in preparing students for a professional medical career. They carried out a Scoping review to determine the elements of effective learning behavior for PBL using the protocol by Arksey and O'Malley [41]. The protocol identifies the research question, selected relevant studies, collected data, and collated, summarized, and reported results. The researchers found three categories of elements that proved effective in the achievement of learning outcomes in PBL. One category is intrinsic empowerment which includes pro-activity, organization, diligence, and resourcefulness. Another is called entrustment which considers students as assessors, students as teachers, feedback-giving, and feedback-receiving. The functional skills theme contains time management, digital proficiency, data management, and collaboration.

Dolmans' et al. findings should be pointed out [40]. They discovered two issues related to the implementation of PBL. One is dominant facilitators, and the other is dysfunctional PBL groups. These problems inhibit students' self-directed learning and reduce their satisfaction level with the PBL session. Therefore, it is essential to have qualified facilitators.

The teaching method of PBL is important for engineering education, especially for engineering design. It has been used to carry out many successful experiments in the real world. The authors of this chapter are researchers who decided to carry out similar activities in the virtual environment of Second Life (SL) because of the benefits it provides. Students can work from anywhere in the world, at any time, and at their own pace. Also, they prepare and use primitives (3-D virtual objects of various sizes, shapes, color, etc.) for their building projects. Therefore, expensive building materials are not required, and potential safety risks are avoided. The next section presents and describes some of the authors' successful PBL activities in SL.

4.2 Problem-Based Learning (PBL) Activities in Second Life (SL)

The authors are researchers who carried out successful PBL activities with students in the virtual environment of SL. The projects took place in virtual buildings with virtual classrooms on Japan's island (in SL) that is owned by Nagaoka University of Technology. The virtual classrooms included red chairs, tables, a podium, and white boards (for posters and power point presentations) which provided a sense of reality (see Fig. 4.1) [43]. Speech and chat functions were available for communication and an area outside of the classroom (the sandbox) was used for building projects. Student teams for these projects included three to five participants, mostly between the ages of 16 and 18 years old. PBL presented the students with challenging problems that related to their daily lives. The students received guidance from their instructors and worked cooperatively to find solutions to the problems. To start, the instructors and students registered and named an avatar (virtual person) to carry out activities on behalf of them. They made their avatars move by using tasks such as

Fig. 4.1 An instructor's avatar presents a lecture to students in a virtual classroom on Japan's island in SL

Fig. 4.2 An instructor watches students prepare primitives (3-D virtual objects) for their building projects in the sandbox area

walking, running, and flying. The teleport function was used to transport their avatars to different locations in SL. Team members participated in brainstorming and decision-making sessions to obtain possible solutions to the problems that needed to be solved. The participants also designed and prepared primitives (three-dimensional virtual objects like cubes of various sizes, shapes, color, texture, etc.) for their building projects which took place in the sandbox area (see Fig. 4.2) [43]. Before taking part in building projects, the students obtained information about this task and practiced making primitives at various locations in SL like the Ivory Tower. In SL, expensive building materials are not necessary and potential safety risks are avoided. In addition to solving problems, the students completed surveys to provide more information about their projects.

STUDENTS DESIGN AND BUILD THE HOUSE OF THE FUTURE DURING THE GLOBAL WARMING ERA

In a specific project conducted in SL, teams from the United States and Japan were tasked with addressing the following challenge: What would a typical house look like in an era of global warming? Both teams approached the activity enthusiastically and successfully completed their designs. The U.S. team created a dome-shaped structure with a floor composed of synthetic wood to conserve Earth's natural forests. Their house also featured a roof equipped with solar panels (see Fig. 4.3) [44]. Meanwhile, the Japanese team designed an energy-efficient, dome-shaped house with a ceiling that could automatically open to allow cool breezes to circulate (see Fig. 4.4) [44].

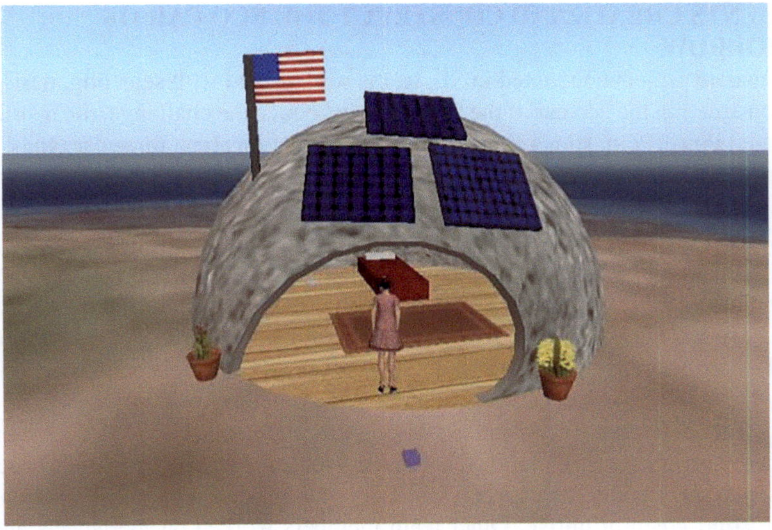

Fig. 4.3 This illustration showcases the house designed by the U.S. team, featuring a dome shape and solar panels in blue

Fig. 4.4 Above is a display of Japan's dome-shaped house

STUDENTS CREATE AND CONSTRUCT THE ECO CAR OF TOMORROW

In a separate project conducted in SL, teams were tasked with selecting, designing, and constructing the eco car of the future. To introduce the challenge, the instructors presented PowerPoint slides containing key information about three car types: solar cars, nuclear cars, and fuel cell cars. The U.S. team was instructed to choose one of these options. Within the virtual classroom, they discussed and compared the car types based on energy efficiency, environmental sustainability, and safety. Safety was a primary concern for the students, leading them to conclude that the solar car was the most viable option. They noted that the solar car is powered by sunlight (a readily available energy source) was both efficient and practical. The other car types raised safety concerns; for instance, nuclear cars posed risks related to radiation and waste disposal, while fuel cell cars had issues with hydrogen flammability.

The team collaboratively sketched a design for their solar car, with each member assigned specific parts to build. One avatar created the small passenger section, represented by a green hemisphere, while another constructed the car's body using a long, solid green primitive. Other team members worked on the wheels and the solar panels, which were depicted as flat, blue primitives. Figure 4.5 displays the completed eco car designed by the U.S. team [45].

Similarly, Japan's team followed an analogous presentation and design process. They created two eco cars: one resembled a boat with wheels and a sail harnessing wind energy for motion, and the other featured propellers and a cigar-shaped body. Survey results revealed that both the U.S. and Japanese teams encountered challenges with communication, avatar navigation, and the creation of primitives. Despite these difficulties, the students found the project enjoyable and expressed interest in participating in similar activities in the future.

Fig. 4.5 The U.S. instructor evaluates the eco car designed by her team

U.S. STUDENTS UNDERTAKE A PBL PROJECT ON NUCLEAR SAFETY IN SL

Using the voice (speech) function in SL, the U.S. instructor, Barry, guided her students through a nuclear safety project. They convened in a virtual classroom, where Barry delivered a forty-minute Power Point presentation. The lecture covered several key points. She began by explaining the significance of nuclear energy and proceeded to discuss the disaster at Japan's Fukushima power plants, which was triggered by a massive earthquake and tsunami in March 2011. Barry informed her students that similar risks could occur in the U.S. and other nations that depend on nuclear energy.

She then addressed types of radiation and their levels of penetration (see Fig. 4.6) [46]. Also, she shared insights into daily radiation exposure from sources such as space, food, and other environmental factors. A crucial part of the lecture focused on radiation protection. Key points included: (1) the necessity of a shield between individuals and radiation sources, (2) the importance of the materials used for shielding, (3) the reduction of dose rates with increased distance, as the dose rate is inversely proportional to the square of the distance, and (4) the reduction of dose volume by minimizing exposure time. Barry highlighted the critical role of shield materials and provided the students with an equation to compute shielding capabilities (see Fig. 4.7) [46]).

The group worked through a sample calculation together in the virtual classroom. They then tackled the following question: Which metals are most effective at shielding humans from radiation? The students calculated and compared the

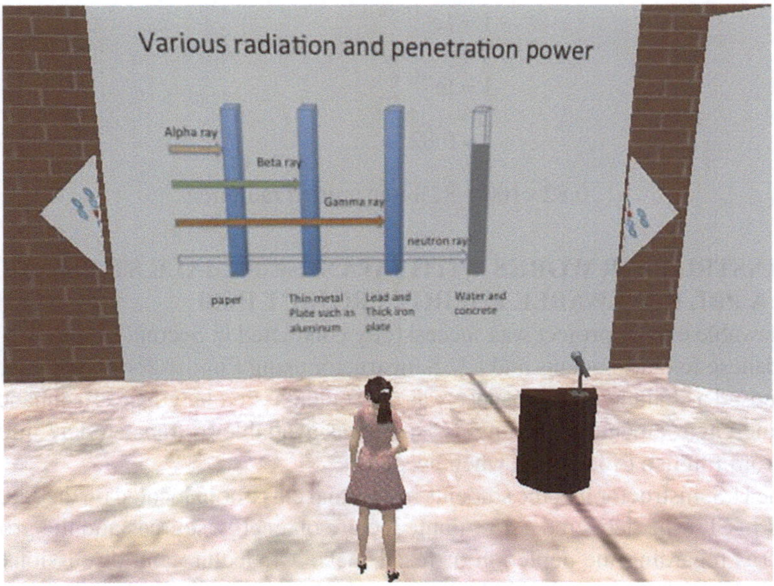

Fig. 4.6 The U.S. instructor uses a Power Point presentation to discuss radiation types with her students

I = Io e^{-ub}
where I (the letter) is transmitted X-ray Intensity,
I subzero is an incident X-ray Intensity, mu is a decay coefficient, and b is the thickness of shielding
(cm)

Fig. 4.7 An equation is provided to calculate the percentage of transmitted radiation

shielding effects of copper and aluminum, while also examining the impact of thickness on shielding effectiveness. They discovered that copper offers better protection than aluminum and mathematically demonstrated that thicker shields significantly reduce radiation penetration. The students concluded that nuclear energy can be safe if shield materials are chosen carefully and are of sufficient thickness to lower radiation penetration to acceptable levels. This PBL activity left the U.S. group confident in their ability to determine safe shielding requirements by applying equations and consulting specialized handbooks containing decay coefficients of various materials.

Sample Problem Determine the percent of transmitted radiation through a copper shield that is 0.1 cm thick. The decay coefficient for copper is 1.98 and I_0 is 1.0.

Substitute these values into the equation and solve for I, Next, multiply the final answer by 100 to obtain the percent of transmitted radiation.

$$I = I_0 e^{-ub}$$

$$I = 1e^{-1.98(0.1)}$$

$$I = 1e^{-0.198}$$

$$I = 0.82$$

$$0.82 \times 100 = 82\% \text{transmitted radiation}$$

U.S. INSTRUCTOR WORKS WITH JAPANESE FEMALE STUDENTS FOR A PBL RENEWABLE ENERGY PROJECT IN SL

A renewable energy project was successfully conducted in Second Life with a team of Japanese female students and a U.S. instructor using English for communication. This posed a challenge, as Japanese was the students' first language. To enhance the learning environment, a teaching assistant was available to translate key information from English to Japanese when needed.

The U.S. instructor met the students in a virtual classroom located on the Japanese island in SL. She delivered a 45-minute lecture on energy sources and the growing concerns regarding the depletion of fossil fuels. The lecture and discussions took place in English through the chat function, which operates similarly to text messaging. Given the language barrier, the teaching assistant provided translation

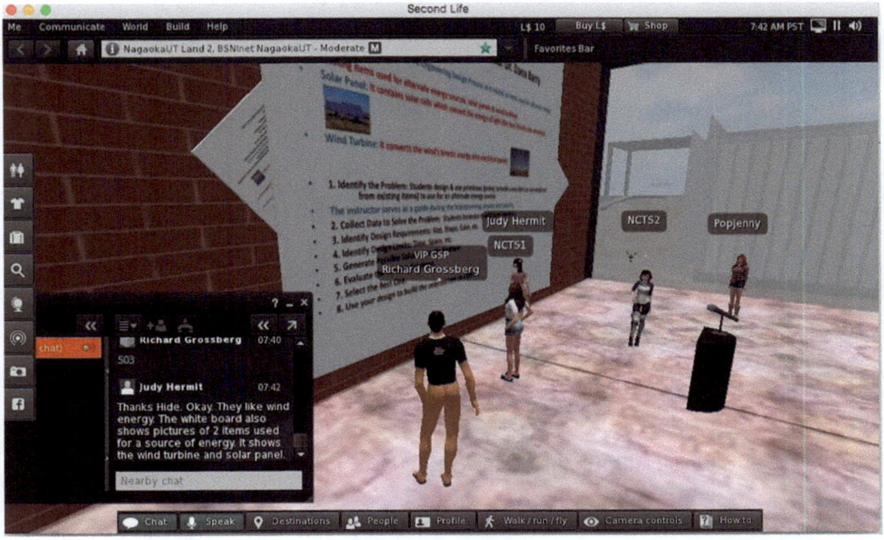

Fig. 4.8 Students review a poster presenting information about renewable energy in their virtual classroom

assistance when necessary. A poster displayed general information about energy sources and outlined the renewable energy challenge alongside a problem-solving model. Figure 4.8 shows students examining the poster [47].

The students were tasked with selecting a type of renewable energy and designing and building a related item using primitives (3-D virtual objects such as cubes). They began by engaging in brainstorming sessions to explore potential solutions. The instructor acted as a guide, while the teaching assistant offered translation support as needed. The students showed enthusiasm and chose wind energy as their renewable energy source, planning to create a specialized type of airplane for its application.

The girls were given general instructions on how to prepare and use primitives. They diligently practiced these skills independently to gain proficiency. Additionally, they used software available in SL to further refine their ability to create primitives. Working as a team, the group collaborated to construct their plane. One student was responsible for the body, another crafted the propellers, and a third designed the wings, among other components. They also coordinated to select specific colors for various parts of the plane. Their airplane functioned similarly to a wind turbine, featuring a standard propeller at the front and a distinctive one at the rear. The wind powered the propellers, which rotated a rotor connected to a main shaft. This shaft spun a generator to produce electricity, enabling the plane to function. Their completed airplane is shown in Fig. 4.9 [47].

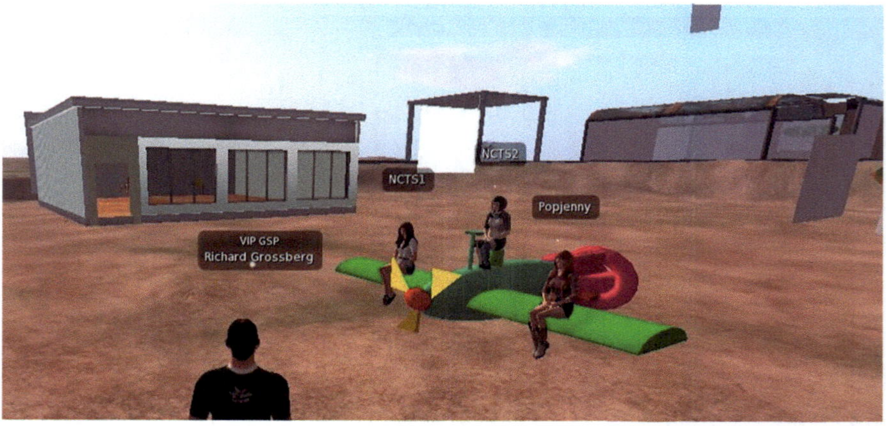

Fig. 4.9 The girls proudly display their completed airplane in the sandbox area

ELEMENTARY SCHOOL TEACHERS PREPARE AND SHARE CREATIVE LESSONS IN SL

A virtual workshop focusing on creative teaching was conducted for elementary school teachers of young children. This initiative was led by the authors to address the growing need for qualified STEM (science, technology, engineering, mathematics) graduates. The United States and other nations require an adequate number of STEM graduates to remain competitive globally, creatively solve complex problems, and supply future scientists, engineers, and innovators. As a result, it is crucial to have motivated educators who can inspire and encourage students of all ages to pursue academic studies, particularly in STEM fields. The authors, who had already delivered numerous lessons to students in virtual classrooms, decided to use a similar approach for this workshop targeting teachers.

The workshop was held in Second Life on an island managed by Nagaoka University of Technology in Japan. The participants, many of whom had no prior experience with virtual environments, greatly enjoyed the workshop and appreciated the creative atmosphere of the virtual classroom. The enthusiastic instructor introduced five innovative teaching approaches for the participants to explore, discuss, and apply in their classrooms successfully. Figure 4.10 showcases a poster highlighting some of these teaching techniques [48].

The creative teaching strategies included reverse engineering, multisensory lessons, student interest projects, reading and solving a mystery, and innovative e-learning. Reverse engineering involves dismantling an object to study its individual components (such as examining an orange). Multisensory lessons engage the senses of sight, smell, touch, hearing, and taste to accommodate diverse learning styles. Some students learn best through visual input, while others excel through auditory cues, etc. For student interest projects, teachers guide students to select a topic they are passionate about, then pose a related question for research. For instance, if a student is interested in food, they might conduct a project exploring

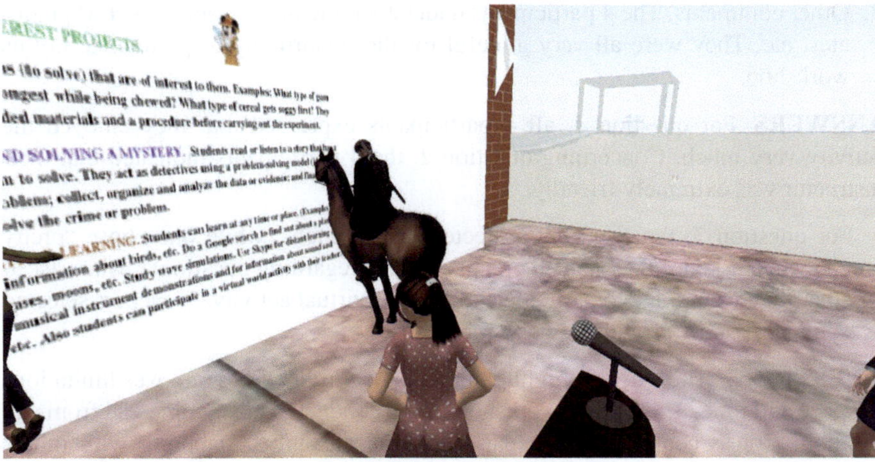

Fig. 4.10 A teacher, mounted on a horse, carefully reads about the creative teaching approach for solving a mystery

different aspects of it. In the reading and solving a mystery activity, students read or listen to a story involving a problem or crime. Acting as detectives, they use a problem-solving model to identify the issue, collect, organize, and analyze evidence, to make the best decision for resolving the problem. The innovative e-learning approach employs creative uses of software and digital media, enabling students to learn at anytime and anywhere. This is particularly beneficial for activities that are impractical or impossible to perform in real life.

After the five methods were presented and discussed, the teachers were asked to select one approach and develop an engaging lesson to share with both their students and the workshop participants. To gather additional feedback, each teacher was requested to complete a survey provided during the workshop.

SURVEY

(Circle or underline the letter of your answer for questions 1–4. Then answer questions 5–7.)

1. Did you enjoy the workshop? A. Very much. B. Pretty much. C. Neutral. D. Not so much. E. Not at all.
2. Was the instructor friendly to you? A. Very much. B. Pretty much. C. Neutral. D. Not so much. E. Not at all.
3. Did you get creative ideas to use with your students? A. Very much. B. Pretty much. C. Neutral. D. Not so much. E. Not at all.
4. Do you want to carry out another virtual activity? A. Very much. B. Pretty much. C. Neutral. D. Not so much. E. Not at all.
5. How would you compare a workshop in the real world to one in the virtual world? (What advantages are available in the virtual world?)
6. What teaching approach did you select for your creative lesson?

7. Other comments. The 4 participants understood the importance of STEM graduates, etc. They were all very grateful for the opportunity to participate in this workshop.

ANSWERS For question 1, all 4 participants expressed that they enjoyed the activity very much. Concerning question 2, the 4 participants indicated that their instructor was extremely friendly.

For question 3, three teachers selected "very much," while one chose "pretty much" when asked if they gained creative ideas. Regarding question 4, when asked whether they would like to participate in another virtual activity, 3 teachers selected "very much," and 1 provided a neutral response.

QUESTION 5 The teachers highlighted that the virtual world removes limitations of space and time. This allows workshops to be conducted at any time and from any location. Consequently, teachers can communicate and collaborate with others regardless of their physical location. The virtual environment eliminates the need for costly building materials and mitigates potential safety hazards that may arise in real-life scenarios.

QUESTION 6 Two examples were provided. For a student interest project, a teacher might pose the following question to the students: "What will your school look like in the future?" Each student would then select an item, such as their desk, and design it for the future. Another example involved a teacher working with very young students, who used the innovative e-learning method to teach about animals around the world. The students chose their favorite animals, searched for images online, and took virtual tours to the habitats (such as countries or continents) of their chosen animals.

4.3 Conclusions

This chapter introduces the teaching methodology known as problem-based learning (PBL) and explores the innovative ways it was implemented to conduct effective activities in virtual classrooms located on a Japanese island in Second Life (SL). PBL presents students with complex, open-ended problems related to real-life situations. Under the guidance of their instructors, students collaborate in groups and engage in brainstorming sessions to devise solutions to these challenges. This instructional approach is particularly significant in engineering education, especially in the context of engineering design, and has been successfully applied in numerous real-world experiments.

The authors of this chapter, as researchers, opted to replicate these activities in the virtual setting of SL due to its numerous advantages. It enables students to participate from anywhere globally, at any time, and at their own pace. Moreover, students utilize primitives (three-dimensional virtual objects of varying sizes, shapes, and colors) for their construction tasks, eliminating the need for costly building materials and mitigating safety risks.

This chapter outlines and examines several successful PBL initiatives carried out by the authors in SL. For instance, one activity involved students designing and constructing a futuristic house suitable for the global warming era. Another project tasked students with creating an eco-car that is energy-efficient, safe, and environmentally sustainable. A third activity focused on nuclear safety, where students investigated which types of metals can effectively shield humans from radiation. In a renewable energy project, students designed a specialized airplane powered by wind energy. Lastly, a virtual workshop was conducted for elementary school teachers (educators of young children), introducing them to creative teaching methods to inspire and engage their students. The teachers selected one of these techniques, developed a creative activity for their students, and shared it with participants in the SL workshop.

References

1. Wang, C.C., Zhao, L.: The process of implementing problem-based learning in a teacher education program: an exploratory case study. Cogent Educ. **8**(1) (2021). https://doi.org/10.1080/2331186X.2021.1996870
2. Barrows, H., Tamblyn, R.: Problem-based learning: an approach to medical education. Springer (1980)
3. Blackburn, G.: A university's strategic adoption process of an PBL-aligned eLearning environment: an exploratory case study. Educ. Technol. Res. Dev. **65**(1), 147–176 (2017). https://doi.org/10.1007/s11423-016-9472-3
4. Chang, D.R., Lin, M.C.: A study of the effects and limitations of the application of problem-based learning on a student teaching curriculum. J. Teach. Educ. Professional Dev. **9**(2), 1–26 (2016)
5. Chang, N.C., Hsu, H.Y.: A study on integrating problem-based learning into the innovative teaching in information literacy and ethics. J. Educ. Media Libr. **53**(2), 171–209 (2016). https://doi.org/10.6120/JoEMLS.2016.532/0010.RS.CM
6. Chang, W.J., Yeh, Z.M.: A study of the effectiveness of problem-based gender relationship and communication by applying interactive e-learning to assist high school and vocational high school students. J. Res. Educ. Sci. **54**(4), 85–114 (2009)
7. Chen, F.R.: Application of problem-based learning in counseling learning for college students. J. Educ. Res. **173**, 44–52 (2008)
8. Chen, K.T., Wang, B.Y.: A study on the teaching experiment of problem-based learning on materials and methods of teaching mandarin at elementary school teacher education. J. Res. Element. Educ. **1**, 163–192 (2005)
9. Dolmans, D.H.J.M., Loyens, S.M.M., Marcq, H., Gijbels, D.: Deep and surface learning in problem-based learning: a review of the literature. Adv. Health Sci. Educ. **21**(5), 1087–1112 (2016). https://doi.org/10.1007/s10459-015-9645-6
10. Edward, S., Hammer, M.: Laura's story: using problem-based learning in early childhood and primary teacher education. Teach. Teach. Educ. **22**(4), 465–477 (2006). https://doi.org/10.1016/j.tate.2005.11.010
11. Erdogan, T., Senemoglu, N.: PBL in teacher education: its effects on achievement and self-regulation. High. Educ. Res. Dev. **36**(6), 1152–1165 (2017). https://doi.org/10.1080/07294360.2017.1303458

12. Ertmer, P.A., Simons, K.D.: Jumping the PBL implementation hurdle: supporting the efforts of K-12 teachers. Interdiscip. J. Prob. Based Learn. **1**(1), 5 (2006). https://doi.org/10.7771/1541-5015.1005
13. Gallagher, S.A.: The role of problem-based learning in developing creative expertise. Asia Pac. Educ. Rev. **16**(2), 225–235 (2015). https://doi.org/10.1007/s12564-015-9367-8
14. Golightly, A., Raath, S.: Problem-based learning to foster deep learning in preservice geography teacher education. J. Geogr. **114**(2), 58–68 (2015)
15. Gunter, T., Alpat, S.K.: The effects of problem-based learning (PBL) on the academic achievement of students studying 'Electrochemistry'. Chem. Educ. Res. Pract. **18**(1), 78–98 (2017). https://doi.org/10.1039/C6RP00176A
16. He, Y.F., Du, X.Y., Toft, E., Zhang, X.L., Qu, B., Shi, J.N., Zhang, H., Zhang, H.: A comparison between the effectiveness of PBL and LBL on improving problem-solving abilities of medical students using questioning. Innov. Educ. Teach. Int. **55**(1), 44–54 (2018). https://doi.org/10.1080/14703297.2017.1290539
17. Helmi, S.A., Mohd-Yusof, K., Phang, F.A.: Enhancement of team-based problem-solving skills in engineering students through cooperative problem-based learning. Int. J. Eng. Educ. **32**(6), 2401–2414 (2016)
18. Hmelo-Silver, C.E.: Problem-based learning: what and how do students learn? Educ. Psychol. Rev. **16**(3), 235–266 (2004)
19. Hsu, C.H.: Case study on applying problem-based learning to the student teaching curriculum. J. Res. Educ. Sci. **58**(2), 91–121 (2013). https://doi.org/10.3966/2073753X2013065802004
20. Loyens, S.M.M., Magda, J., Rikers, R.M.J.P.: Self-directed learning in problem-based learning and its relationships with self-regulated learning. Educ. Psychol. Rev. **20**(4), 411–427 (2008). https://doi.org/10.1007/s10648-008-9082-7
21. Yang, K.Y., Chang-Lai, M.L.: The theoretical background and teaching process of problem-based learning. Chung Yuan J. **33**(2), 215–235 (2005). https://doi.org/10.6358/JCYU.200506.0215
22. Zhou, C.F., Shi, J.N.: A cross-cultural perspective to creativity in engineering education in problem-based learning (PBL) between Denmark and China. Int. J. Eng. Educ. **31**(1A), 12–22 (2015)
23. Yew, E., Goh, K.: Problem-based learning: an overview of its process and impact on learning. Health Profession Educ. **2**(2), 75–79 (2016)
24. Ghani, A.S.A., Rahim, A.F.A., Yusoff, M.S.B., Hadie, S.N.H.: Effective learning behavior in problem-based learning; a scoping review. Med. Sci. Educ. **31**(3), 1199–1211 (2021)
25. Taylor, D., Miflin, B.: Problem-based learning: where are we now? Med. Teach. Taylor Francis. **30**(8), 742–763 (2008)
26. Rakhudu, M.A.: Use of problem-based scenarios to prepare nursing students to address quality improvement in health care unit. Int. J. Educ. Sci. **10**(1), 72–80 (2015)
27. Radcliffe, P., Kumar, D.: Is problem-based learning suitable for engineering? Austr. J. Eng. Educ. **21**(2), 81–88 (2016). https://doi.org/10.1080/22054952.2017.1351131
28. Choi, E., Lindquist, R., Song, Y.: Effects of problem-based learning vs. traditional lecture on Korean nursing students' critical thinking, problem-solving, and self-directed learning. Nurse Educ. Today. **34**(1), 52–56 (2014)
29. Wong, D.K.P., Lam, D.O.B.: Problem-based learning in social work: a study of student learning outcomes. Res. Soc. Work. Pract. **17**(1), 55–56 (2007). https://doi.org/10.1177/1049731506293364
30. Wijnen, M., Loyens, S.M.M., Smeets, G., Kroeze, M., van der Molen, H.: Comparing problem-based learning students to students in a lecture-based curriculum: learning strategies and the relation with self-study time. Eur. J. Psychol. Educ. **32**(3), 431–447 (2017)
31. Armitage, A., Pihl, O., Ryberg, T.: PBL and creative processes. J. Prob. Based Learn. Higher Educ. **3**(1) (2015)
32. Harun, N.F., Yusof, K.M., Jamaludin, M.Z., Hassan, S.A.H.S.: Motivation in problem-based learning implementation. Procedia Soc. Behav. Sci. **56**, 233–242 (2012)

33. Smith, G.F.: Problem-based learning: can it improve managerial thinking? J. Manag. Educ. **29**(2), 357–378 (2005). https://doi.org/10.1177/1052562904269642
34. Siew, N.M., Mapeala, R.: The effects of problem-based learning with thinking maps on fifth graders' science critical thinking. J. Balt. Sci. Educ. **15**(5), 602 (2016)
35. Li, H.C., Stylianides, A.J.: An examination of the roles of the teacher and students during a problem-based learning intervention: lessons learned from a study in a Taiwanese primary mathematics classroom. Interact. Learn. Environ. **26**(1), 106–117 (2018). https://doi.org/10.1080/10494820.2017.1283333
36. Wilder, S.: Impact of problem-based learning on academic achievement in high school: a systematic review. Educ. Rev. **67**(4), 414–435 (2015). https://doi.org/10.1080/00131911.2014.974511
37. Gutman, M.: The influence of problem-based learning communities on research literacy and achievement goal motivation. Int. J. Educ. **6**(4), 31–41 (2018)
38. Tarhan, L., Ayyildiz, Y.: The views of undergraduates about problem-based learning applications in a biochemistry course. J. Biol. Educ. **49**(2), 116–126 (2015). https://doi.org/10.1080/00219266.2014.888364
39. Kartamiharja, M.R., Sopandi, W., Anggraeni, D.: Implementation of problem-based learning (PBL) approach in chemistry instructional with context of tofu liquid waste treatment. Int. J. Educ. Res. **19**(5), 47–77 (2020)
40. Dolmans, D., Grave, W., Wolfhagen, I., Van der Vleuten, C.: Problem-based learning: future challenges for educational practice and research. Med. Educ. **39**(7), 732–741 (2005)
41. Arksey, H., O'Malley, L.: Scoping studies: towards a methodological framework. Int. J. Soc. Res. Methodol. **8**(1), 19–32 (2005). https://doi.org/10.1080/1364557032000119616
42. Parikh, A., McReelis, K., Hodges, B.: Student feedback in problem-based learning: a survey of 103 final year students across five Ontario medical schools. Med. Educ. **35**(7), 632–636 (2001)
43. Barry, D.M., Kanematsu, H., Ogawa, N., McGrath, P.: Technologies for teaching during a pandemic. Proc. Comput. Sci. **192**, 1583–1590 (2021)
44. Barry, D.M., Kanematsu, H.: Virtual reality enhances active student learning. Proc. Comput. Sci. **207**, 408–415 (2022)
45. Kanematsu, H., Kobayashi, T., Ogawa, N., Barry, D.M., Fukumura, Y., Nagai, H.: Eco car project for Japan students as a virtual PBL class. Proc. Comput. Sci. **22**, 828–835 (2013)
46. Barry, D.M., Kanematsu, H., Fukumura, Y., Kobayashi, T., Ogawa, N., Nagai, H.: U.S. students carry out nuclear safety project in a virtual environment. Proc. Comput. Sci. **22**, 1354–1360 (2013)
47. Barry, D.M., Kanematsu, H., Lawson, M., Nakahira, K., Ogawa, N.: Virtual STEM activity for renewable energy. Proc. Comput. Sci. **112**, 946–955 (2017)
48. Barry, D.M., Kanematsu, H., Nakahira, K., Ogawa, N.: Virtual workshop for creative teaching of STEM courses. Proc. Comput. Sci. **126**, 927–936 (2018)

Chapter 5
Building Virtual Worlds for the Video Game Industry

Morgan Hastings

Abstract Virtual worlds are computer-based simulated environments where users can interact with others by using an avatar or playable character. Several graphical applications are presented and described. Game design is the creation and shaping of the rules and content of a game. It includes the creation of stories, characters, goals, and challenges. It can be used for entertainment, education, and other purposes. Developing the graphics for VR worlds used in video games involves understanding how to tell a story through a mostly static environment. Different styles can be employed to represent the world: realistic, stylized or cartoon-based, abstract or artistic. There are special elements for world building in video games, movies, and literature: geology, flora, fauna, sociology, and physics. Developing worlds for video games involves a modeling package. Such a package is typically Autodesk Maya, Substance Painter and/or Photoshop for texturing, and a game engine that is either proprietary or commercial such as Unreal Engine or Unity. Immersion into this virtual world is the goal. Therefore, one needs to eliminate or reduce anything that might distract the player from suspension of disbelief.

Keywords Virtual worlds · Game design · Avatars · Playable characters · Modeling package for game design · Unreal engine · Immersion into virtual reality (VR) · Video game industry

5.1 Introduction

Immersion in a virtual world is a commonly acknowledged concept, but to grasp its essence, one must first understand what immersion entails. I often illustrate immersion to my students by presenting an image of a Boeing 247 passenger plane from

M. Hastings (✉)
State University of New York at Canton, Canton, NY, USA
e-mail: hastingsm@canton.edu

D. M. Barry, H. Kanematsu (eds.), *Applications of Metaverse and Virtual Reality to Creative Education and Industry*, Intelligent Systems Reference Library 267, https://doi.org/10.1007/978-981-96-3341-8_5

the 1930s and prompting them to envision what they perceive, hear, feel, and even potentially smell in that environment. While virtual reality environments typically lack scent reproduction, save for historical exceptions like the Sensorama mechanical device from the 1950s, they can effectively engage the remaining senses through VR equipment [1].

It's remarkable how extensively the students and I delve into this sensory experience of flying aboard a Boeing 247, the pioneering commercial airliner of the 1930s, during our discussions, often spanning an hour. With its maximum altitude of 11,500 feet, capacity for only ten passengers, air-conditioned cabin, sound-deadening walls, and twin propellers, the Boeing 247 offers a rich sensory tapestry. From feeling the vibrations of the propellers and turbulence to experiencing varying G-forces during acceleration, the tactile sensations would be vivid [2].

In terms of sight, passengers could observe the landscape below due to the limited altitude, as well as fellow passengers, the cockpit without a door, a single flight attendant and overhead cargo. The auditory landscape was dominated by the roaring propellers, despite efforts to dampen sound, along with the mechanical sounds of retracting landing gear, pilot communications, passenger conversations, and the overall ambience of flight.

Considering this level of detail, it becomes evident how challenging it is to craft a VR experience that truly immerses the viewer. We can only do as much as possible with current technology and then rely on human suspension of disbelief to bridge the remaining gaps. Factors that detract from immersion, such as low frame rates or lag, stemming from the limitations of the computer hardware, must also be considered. It's important to note that immersion is not solely dependent on realism; even stylized or cartoonish games can captivate players when their design aligns cohesively with their intended experience, sustaining immersion [3] (Fig. 5.1).

5.2 Imagining a VR World

The information presented here is from my own experiences in the video game industry, where I have been employed as an environment artist since 1995. Additionally, I bring experience as a university lecturer, instructing students in conceptualizing and creating objects and environments. The video games industry is often secretive in their methodology due to the novelty of the techniques as well as employees having non-disclosure agreements. Increasingly, there is much more sharing of technology at conferences and online discussion boards [4].

I categorize worlds into five different key elements which jump-start student's imagination: geography, flora, fauna, sociology and physics. Geography encompasses planetary features, including oceans and mountains, either of alien imagination or something of a mixture more native to earth. Flora pertains to the world of plant-life: trees, flowers, grasses and weeds. Fauna is animal life ranging from insects to mammals or some form of extra-terrestrial creatures. Some worlds can have a crossover of flora and fauna, just as J.R.R. Tolkien's Middle-Earth has "Ents,"

Fig. 5.1 The Boeing 247 aircraft

which are sentient trees that can speak and move and walk about rather slowly. Sociology delves into what a sentient society or group might create or leave as relics in the world. Even in the absence of sentient life, the environment might leave behind a narrative of a history in its relics or written language. Seeing a rock with alien inscriptions in an uninhabited land creates a piece to a puzzle that asks to be investigated. Physics governs the natural laws such as gravity. The environment may have greater or lesser gravity, or, in a magical world, a mixture of both. Physics can also be suspended as in a magical/spiritual world where a player could suspend the natural physics of the world to fly, or the player could encounter a floating land mass suspended by mystical energies.

When considering the narrative of the environment, I endeavor to inspire students to entertain the player with their world and ask themselves, "What about this environment makes it interesting? What can I do to create a static environment that tells a story?" It is one thing to recreate an environment, but to tell a story that piques the player's interest creates a memorable experience. This was something I learned at Lucas Arts while working on the game, "Republic Commando" [5]. I moved beyond merely modeling and designing a space but entertained the user to be curious about the use of the space and its narrative [6].

I give this example to my students about creating a narrative from a static scene. Imagine a nice country road with tall grass on either side with a small somewhat abandoned barn off to the side. The scene is peaceful with a nice tree in front and a quaint little painted wooden sign reading, "Welcome." As we step closer, we see a bloody handprint on the sign and bloody bare footprints leading out of the barn crossing into a wooded area on the other side of the dirt road.

Suddenly, our idyllic scene changes tone to a horrifically dreadful one; our minds have created suspense with a subtle suggestion. No longer is this scene innocent but rather ominous. We either turn around and go home quickly or trudge forth in fearful anticipation of what we might encounter. Getting the player to imagine (without literally describing) is a key element to world-building akin to techniques in literature and cinema. Alfred Hitchcock was a master of suspense and could arouse fear at what may occur with what might appear completely mundane in a different setting [7].

Before my students embark on writing about and creating their own worlds, we study and write about worlds in movies, literature, and television programs that they might find captivating. We study Tolkien's Middle-Earth, Harry Potter, Star Trek, Star Wars, the Alien series, Max Max and other complex worlds.

There are many worlds in Star Trek and Star Wars each with their own version of the five elements. In "Mad Mad: Fury Road" the world is post-apocalyptic dominated by stylized vehicles. While the movie was still scripted, it relied heavily on storyboarding which infused it with a dynamic and exciting energy making a compelling visual spectacle. Harry Potter's world has suspended physics due to the magical narrative. The Alien series draws upon the macabre and surreal visionary conceptual artwork of H.R. Giger [8].

Any world can be created in VR, but what the designer should strive for is to make it compelling and interesting for players to explore and to inspire wonder or moods. Worlds need not always be realistic in nature; even those with a stylized, cartoonish, or abstract aesthetic can be profoundly immersive. The Rick and Morty VR game is a humorous and cartoonish world that was very well received. Sony's PSVR game, Astro Bot, is a fantastic example of a cartoonish world that is extremely captivating. So long as the environment remains consistent with its own style and quality it will most likely be engaging.

5.3 Concept Art

Concept art plays a pivotal role for world-building and helps to lighten the load of the 3D artists so they can focus on bringing the whole world together technically and artistically. This doesn't mean that 3D artists no longer need to imagine and create, as there is still plenty of need for their skills. Many large high profile game companies employ dedicated 2D artists to assist in the creation pipeline [9].

In my VR Worlds class, students first create a "Sandbox" world to familiarize themselves with the applications. They start by creating landscapes and importing pre-existing objects and create materials for each. This hands-on approach helps them to understand the tools and their capability for immersion.

Once this process is complete, students move on to conceptualizing their world using the five world building elements in a proposal paper containing their concept art. They can include photographs of areas that already exist, their own illustrations or AI generated artwork.

If a student has a world that closely resembles earth, they must add a subtle narrative to it that hints at a story awaiting discovery. For instance, the player might find ruins in a forested area with foreign inscriptions and surrounding particle systems giving the impression of enigmatic alien origin.

5.4 Technology

The technology and applications used to create VR worlds can vary depending on the company. For modeling objects, characters, and modular components as well as animations, most video game companies use Autodesk Maya, a 3D modeling package widely used in games, movies, and television programs to create objects and architecture. Characters and organic shapes like tree trunks are often created in Zbrush which is a popular 3D sculpting tool. These objects and animations are then imported into a video game engine and arranged [10].

Game engines are where the game is coded, and the graphics and audio and other effects are combined. Many game companies create their own proprietary engines which can save them from paying significant royalties to the companies that produce game engines. However, creating a game engine from scratch can be a very time-consuming and costly process.

Two engines are the most popular: Epic Games' Unreal Engine and Unity [11]. My professional experience has been with Unreal. In 2005, at Lucas Arts, I worked with an early version of the Unreal Engine to make environments. Following Lucas Arts, I used the Unreal Engine to create environments for America's Army which was developed by Secret Level and originally by the U.S. Military.

Textures for the objects and world are created in Photoshop or Substance Painter. While Photoshop is a 2D painting tool, Substance Painter offers the capability to draw directly on the 3D object and create complex layered materials. The choice is usually the artist's preference. However, Photoshop is the default choice for painting and is especially good at creating textures that tile with no seams [12].

Computers used in creating VR worlds are typically PCs with a significant amount of processing power, memory, and storage. Creating environments generates vast amounts of data necessitating rapid processing capabilities. Game development computers must also have state-of-the-art GPUs or Graphical Processing Units. Currently, Nvidia RTX 4x cards are quite popular. GPUs process data in a parallel fashion to free up time for the CPU to handle other essential tasks. The newer GPUs can process real time ray-tracing shadows which is a severely intensive computational task [13].

There are several types of VR HMDs (head mounted displays) that can be used in VR world building (including devices like the HTC Vive, Meta Quest, or Valve Index). Most console game platforms such as the Sony Play Station and Microsoft Xbox are powerful devices and can handle complex graphics, yet may not match the processing power of a desktop PC with a powerful graphics card. As a result, some changes to the environment must be made to ensure optimal performance on the intended platform [14].

5.5 Textures

Textures are 2D images that are applied on 3D objects. Without textures, objects would mostly appear to be a solid color. The texture adds a great amount of detail without increasing polygon count. Textures are applied onto objects using a process called, "UVing." U and V denote the horizontal and vertical positions within 2D space.

There are many ways to conceptualize the UVing of an object. One is unfolding a cube wrapped around another cube. Another is imagining the texture as a label to be applied to a soda can. UVing can also be compared to the process a tailor undertakes with clothes where the cloth of the sleeve is the texture, and the seams correspond directly to the UV seams. The cloth should be smooth and unwrinkled. Also, the UVs should not be stretched and warped.

The three main methods for applying UVs are automatic, cylindrical or normal-based. Normal - based refers to the directions perpendicular to each individual polygon. Maya then applies the texture considering the normal of each polygon. Automatic mapping finds a quick solution to apply the texture but often lacks customization. Cylindrical mapping is just for cylindrical objects.

Using these methods along with "unfolding" the texture, artists can avoid stretching or warping. The texture for a character's head, for example, is applied via a normal-based UV process. Then a seam is cut along the back of the head and subsequently unfolded to equally distribute the UVs about the topology. The 2D texture for a face looks like a very wide stretched-out face. Figure 5.2 shows how UVing a face and unfolding the UVs would look in Autodesk Maya [15].

Modern texturing involves multiple textures to make one single material. A material is a composition of how these textures create a realistic surface that accurately represents the surface of the object. For instance, an object made of soft black rubber will not be particularly shiny but would appear black underneath and much brighter where it is facing the light source, almost appearing white. This phenomenon is called "specularity." If the surface of the object is polished metal, there will be no large white spot, but rather a small white dot, and the surface of the object will be like a mirror reflecting the surrounding environment. The reflection in the texture is programmatically generated in the game engine from the surrounding environment.

These kinds of textures are referred to as, "PBR" textures, or "physically-based rendering" textures. By employing PBR textures, surfaces can be rendered to the point of photographic realism. Some of these textures include albedo, specularity, normal map, roughness, metallic, ambient occlusion, and subsurface scattering. Opacity is another texture that is used to mask out parts of a polygon that are not to be drawn, which is very useful for foliage.

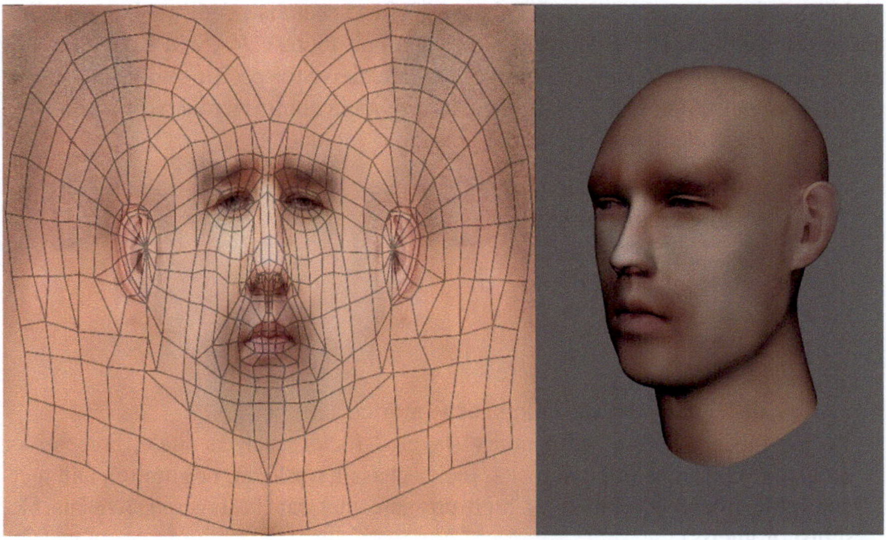

Fig. 5.2 Unfolding UVs on a face to avoid stretching the texture [16]

Albedo, or "diffuse," is the base color of the texture, much like the ink on a label for a bottle. Specularity would show which areas were less shiny or more diffuse. A normal map simulates the bumps or tactile nature of the object. Displacement maps function similarly to normal maps, but alter the surface of the model, tessellating the polygons to make visible protrusions like rocks or bricks. Displacement maps aren't used extensively as using them can take a fair amount of computational power.

Subsurface scattering can be likened to frosted glass. It's not transparent but allows colors to show through. This can also be analogous to holding one of your fingers up to a bright light, seeing a very deep red hue due to the circulating blood.

Ambient occlusion darkens the crevices of an object where the light doesn't easily penetrate. This enhances contrast and detail in a scene, often elevating 3D objects to a higher level of realism [17].

5.6 Planning Outdoor Scenes

At times even the most experienced artist can overlook various elements required for an outdoor scene, leaving it feeling incomplete. To ensure we have included everything to make a scene believable, it is wise to keep a list of elements that will go into the scene being created. References play a crucial role in achieving congruence between the look and feel of the VR world and our perception of reality. Outdoor scenes vary significantly requiring different elements to maintain believability. For instance, a scene on a beach contains distinct elements compared to a forested scene, a snowy landscape, a farm, a desert, or an alien world [18].

In some larger game companies, there are often specialists dedicated to modeling and texturing flora. These companies, or the artists themselves, will have reference books detailing diverse varieties of trees that are native to specific regions. A willow tree is vastly different from a palm tree, as a sequoia contrasts to a redwood.

Some typical elements for outdoor areas may include:

- Grass: wild, tall, dead, cultivated
- Trees: deciduous, coniferous, dead, stumps
- Ivy
- Weeds: tall, ground cover
- Flowers: wild, cultivated
- Bushes: wild, topiaries
- Cattails
- Leaves: different types and colors, new or dried
- Rocks: moss covered, sedimentary (layered rocks, such as shale, breccias or sandstone), igneous (melted rocks, such as basalt, obsidian, or granite), and metamorphic (rocks combined by high pressure that can be layered, like marble, schist or gneiss).

Attention to detail is very important in designing worlds. If an artist creates flora that is not at all native or common for an area this will be noticed and corrected. Many game companies value visual and historical accuracy in the environments they create. Recently I watched a video from an Egyptian historian who was reviewing a video game set in Egypt. Viewers were impressed at the game's level of detail and accuracy. Commitment to accuracy and stylistic congruence helps to suspend the player's disbelief that what they are viewing is a simulation [19].

5.7 Creating Landscapes

Landscapes play a crucial role in creating a VR world with the Unreal Engine and can be easily created, sculpted, painted, and decorated. In the past, making landscapes was fraught with technical challenges especially when texturing them. However, Unreal has largely solved these problems [20] (Fig. 5.3).

Landscapes use a specialized type of material that is multi-layered. Materials in Unreal are "node-based," meaning they have nodes of information that are connected via virtual wires, much like a Moog synthesizer uses patch cables. Nodes serve as a form of visual scripting and can have various customizable functions. In Fig. 5.4, there is a layered node that accepts the input from six different sub-shaders each with their own set of PBR textures. On the left of these node sets is a node that contains the UVs from the landscape, and it is plugged into every sub-shader [21] to ensure proper texturing of the landscape.

This multi-layered material is then used to paint the landscape. There can be various levels of opacity resulting in considerable variety in the visual appearance

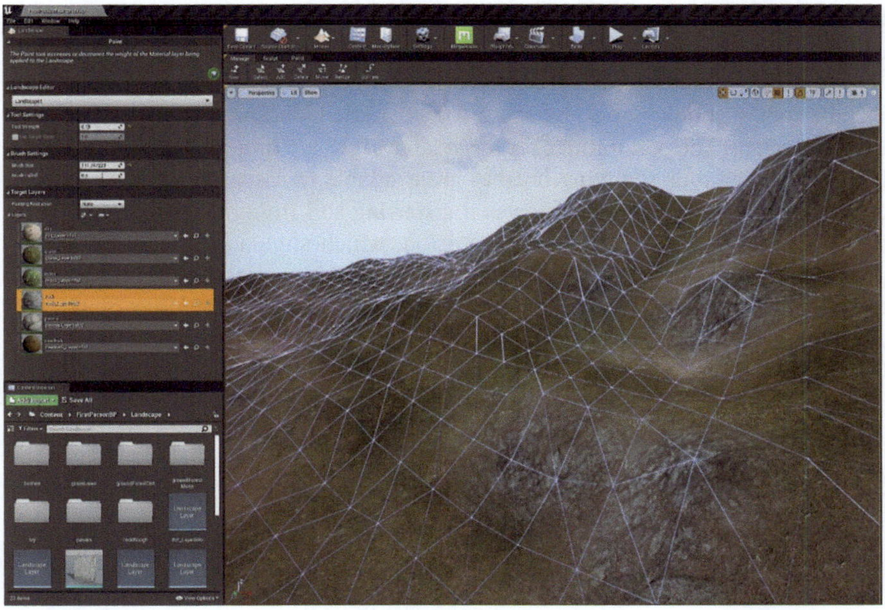

Fig. 5.3 Sculpting and painting a landscape

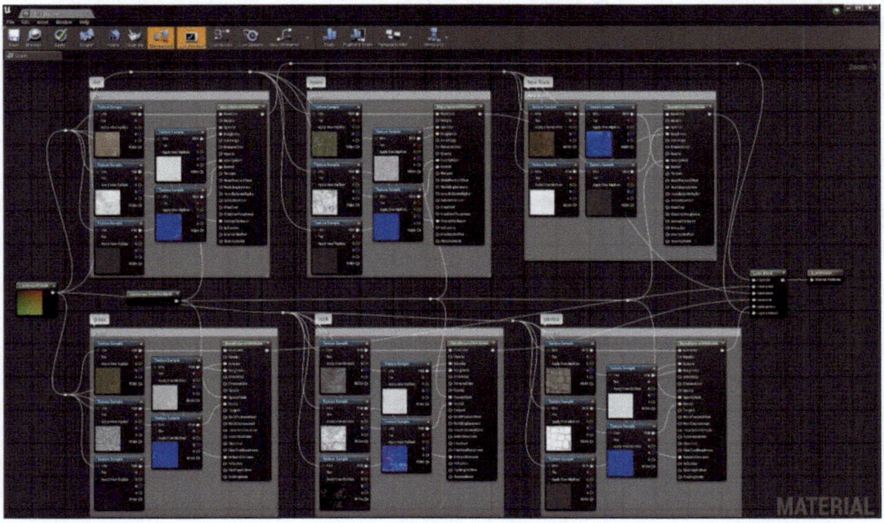

Fig. 5.4 Node system for landscape material

of the ground. Grass can have a bit of dirt, or the rock texture can have a bit of moss growing on it.

Holes can also be created by incorporating a "Landscape Visibility Mask" node as can also be seen in Fig. 5.4. The artist can use these holes to make a tunnel that leads to an underground cavern. Landscapes can be rotated or flipped to make the underside of a floating island.

There is a plethora of landscape types that may be sculpted in Unreal. They include snowy, mountainous, meadow, forest or alien landscapes with very foreign looking terrain. The possibilities for these landscapes are limited only by the artist's imagination. Various sculpting brushes are available to assist the artist in the process, such as erosion which simulates transferring soil from higher to lower elevation over natural change, and hydro erosion, which simulates erosion after long periods of heavy rainfall. Landscape can be flattened or smoothed, and ramps can be created for a gradual transition between elevations. Satellite topology of the earth can also be employed in the form of a grayscale image and used as a heightmap. Once the landscape is sculpted, foliage can be scattered throughout the scene.

5.8 Foliage

The Foliage tool in the Unreal editor is a placement tool for small objects. When the object is designated foliage, it can then be painted on and erased from the landscape easily. This process historically was extremely time consuming, but the Unreal editor solves the problem with this unique tool. Regardless of subsequent sculpting, the foliage tool will move the grass or other object directly on top of the landscape. The foliage can also have very little density (meaning sparsely scattered) or can be very densely packed. The size of the object can also be randomly determined to prevent excessive size uniformity [22].

Foliage in Unreal can be anything ranging from grass, low weeds, tall weeds, tall grass, flowers, and shrubs that one might find scattered along the ground. Even small pebbles and rocks, dried leaves, small sticks, and even trees can be treated as foliage. These components are modeled and UVed in Maya and textured in Photoshop or Substance Painter, imported into Unreal, and scattered. Foliage can also be modeled in another package called, "Zbrush" which is a 3D sculpting tool.

Figure 5.5 shows the scattering tool in action. The white sphere is the "brush" and can be adjusted in both size and density.

Creating the foliage to be scattered is a challenge in video game environments since there are often many small leaves or petals. While the artist could theoretically model all these leaves, it would subsequently increase the time needed to make the environment and substantially increase the computation time for the computer to render. If there are 50 trees in a scene each with 10,000 leaves the frame rate could drop or lag and detract from the immersion of the player's virtual experience.

There are many methods for creating trees. I show my students how to model trees from scratch to avoid dependence on automated processes and specific applications that their company might not use. Automated tree programs like, "Speed Tree" create impressive looking trees that are highly customizable. The artist can make twisted trunks, adjust branch placement and density, and control the number

Fig. 5.5 Painting foliage on a landscape

of leaves. Not all game companies use Speed Tree [23]. However, it is an extremely useful application.

When making trees from scratch, the artist uses a texture to represent groups of leaves to reduce the overall polygon count. Multiple leaves are put on a single texture with an opacity map, which is a black and white image that masks off what isn't drawn. Figure 5.6 demonstrates the diffuse and opacity maps. For grass, a similar type of texture is used.

The trunk is modeled, and major branches emerge from it. Two types of leaf textures are placed around the branches: leaf clumps and small branches. This effectively conveys to the player that the tree is comprised of many leaves. While advancements may be made in the computational power of computers, there will likely always be some degree of "faking" involved in representing trees. Figure 5.7 shows these leaf clumps and tree branches and the fully assembled tree.

5.9 Modular Components

Modular components stand out as perhaps the most crucial technique when it comes to creating architecture for game engines. This process involves identifying patterns in the architecture, modeling each piece, exporting from Maya into Unreal, then assembling them into the final structures. This approach enables quick creation of adaptable architecture that can be expanded for any adjustment in design. It

Fig. 5.6 The two tree leaf textures with opacity maps on the right

streamlines world creation because these individual components can be used repeatedly in any order designed by the artist [24].

In my environment modeling class, I introduce this concept around the midterm and students create pieces to recreate a somewhat abandoned mansion in Poland called "Pałac Luboszyce" (Fig. 5.8) [25]. Using many reference images, students model these individual pieces. Due to the time constraints of the school, the full palace is not modeled; instead, a slightly more simplified version is made. Often some small sacrifices of unique details must be made in favor of facilitating modular component design.

Reducing the complexity of the original palace, the mansion results in just thirteen modular components as can be seen in Fig. 5.9.

Once the building is constructed, to mitigate the obvious repetition of textures, decal "actors" can be used. Decals function by projecting a texture onto surfaces. In the case of the mansion, decals using a texture of dripping water damage are placed over the walls. Additionally, other decals can be added to add diversity to surfaces such as cracks in walls, moss, signs, or graffiti. Decals are also useful to simulate tire tracks or footprints, or even traffic lines on roads. Using a normal map with the decal, it can appear to indent or protrude from the surface.

Fig. 5.7 Creating a tree from the textures in Fig. 5.6

Figure 5.10 shows how the single window modular component can exhibit reduced repetition using decals.

Figure 5.11 shows the completed scene with arranged modular components for landscape, a tree, grass, various foliage, and a small pond decorated with floating water lilies.

Modular construction techniques are also used in internal structures as can be seen in a scene in Fig. 5.12 by student, Zion Vialva. This approach is often called, "hallway systems." Hallway systems can be seen in a variety of games, movies, and television shows. When I worked on the game, "Republic Commando" with Lucas Arts, I created the Republic Assault Ship which used hallway systems for the bridge, hallways leading to it, torpedo rooms, spacecraft hangars and a detention area. They were made up of many small interchangeable parts; straight walls, ascending and

Fig. 5.8 Reference for Pałac Luboszyce. (Credit: Leszekpazderski)

Fig. 5.9 Modular components for Pałac Luboszyce

Fig. 5.10 Decals adding variation to the modular components

descending walls, connecting frames and junctions. The hallways of the Enterprise on Star Trek are another good example of repeated hallway system patterns.

In the space scene in Fig. 5.12, you can see and observe how assembling small components from Fig. 5.13 can create a highly detailed environment. This not only is visually appealing, but also significantly reduces the time required for artists to construct such complicated environments.

5.10 Finishing Touches

Lighting exerts a huge impact on a scene capable of dramatically shaping its ambience. By manipulating the main light source, the shadows can be lengthened or their direction changed. Point lights and spotlights are smaller lights and are commonly used in internal structures. They are often tied to an object to give the appearance of emitting light, such as a flaming torch, candle, or lamp [26].

Particle effects involve phenomena such as fire, steam, or sparks all of which are programmatically controlled. Like the leaves on a branch, particles are displayed on polygons with a transparency map. These particles are then emitted and distributed to simulate real-world effects and can be changed in seemingly countless ways.

Fig. 5.11 Stylized 3D version of Pałac Luboszyce in Unreal 4.27

Fig. 5.12 Space scene using modular components. (Courtesy of Zion Vialva)

Figure 5.14 demonstrates a built-in particle effect of a campfire with extra smoke added. The green arrows indicate in which direction the effect is emitting [27].

Adding volumetric fog can profoundly change the mood of a scene. "Actors" are placed in a scene that controls fog like "Exponential Height Fog" which simulates fog that hovers over the ground. When volumetric fog is turned up, crepuscular rays become visible particularly when an object such as a tree obstructs the light. These

Fig. 5.13 Modular components for the space scene. (Courtesy of Zion Vialva).

Fig. 5.14 Particle effect simulating fire and smoke

Fig. 5.15 Pałac Luboszyce with changed lighting and exponential height fog

light rays are mostly visible in low-light situations. Fog can also obscure the scene's background, enhancing a sense of suspense while simultaneously conserving computational resources as the engine doesn't need to render obscured elements [28].

In Fig. 5.15, the height of the sun is changed, transforming it into a dusk-like evening. Volumetric Fog and exponential height fog are added. What was once an abandoned yet verdant scene, now exudes a heightened sense of suspense.

Sound adds another dimension to the VR experience. Through attenuation, sound can intensify as one approaches the source. For instance, a small pond can have frogs and crickets that grow in volume the closer one gets to it. Additionally, sounds can be triggered by events or actions such as opening a door. One particularly effective use of sound is to link different materials on the landscape to sounds resulting in distinct footstep sounds when the player travels to different terrain [29].

5.11 Conclusion

Creating VR worlds and environments encompasses a wide array of methods which can vary from one artist to another and from one company to another and even between engines. Methods can change very quickly depending on new discoveries in efficiency, or advancement in technology such as a new graphics accelerator or game platform.

Fig. 5.16 Student, Rowan Wysocki, working on his VR world

It is interesting to note that some aspects of environmental creation have not changed very much. We still model objects and use texture maps and lights. While technological advancements have made realistic scenes incredibly lifelike, the downside can be increased production time compared to previous years with less complicated environments (Fig. 5.16).

An environment artist has a multitude of roles: arborist, gardener, architect, painter and sculptor, geologist, writer, and theatrical lighting technician. They must not overlook details many people take for granted in their environment. One missed element can render a scene artificial. However, one slight change of the environment, as simple as changing the intensity of a light, can make a scene suddenly believable. Artists must balance their imagination with their technical skill, uniting them to create immersive environments to entertain, enchant, and delight the player.

References

1. Gould, W.: Boeing (Business in Action), p. 14. Cherry Tree Books, Bath (1995). ISBN 0-7451-5178-7
2. van der Linden, F.R.: The Boeing 247: The First Modern Airliner, p. 1. University of Washington Press, Seattle (1991). ISBN 0-295-97094-4. Retrieved: July 26, 2009
3. https://medium.com/@amysteele1999/immersion-in-gaming-and-entertainment-creating-engaging-virtual-worlds-1ad55e908805
4. https://www.sciencedirect.com/science/article/abs/pii/S0950584921000252
5. https://store.steampowered.com/app/6000/STAR_WARS_Republic_Commando/

6. https://aestranger.com/environmental-storytelling-its-important-for-you/
7. Crisler, B.R.: Hitchcock: Master Melo dramatist. The New York Times. Archived from the original on 12 June 2018 (1938). Retrieved 11 June 2018
8. https://screenrant.com/alien-concept-art-giger-changed-movie/
9. https://www.gamesindustry.biz/a-brief-guide-to-becoming-a-concept-artist
10. https://all3dp.com/2/3d-modeling-for-games-software/
11. https://www.perforce.com/blog/vcs/most-popular-game-engines
12. https://insider.dbsinstitute.ac.uk/the-top-5-softwares-used-by-game-artists
13. https://itsc.ontariotechu.ca/tele/business-and-it/fbit-specs_game-development.php
14. https://www.pcgamer.com/best-vr-headset/
15. https://documents.sessions.edu/eforms/courseware/coursedocuments/maya_i/lesson5.html
16. Model for texture courtesy of composer, Paul Swartzel
17. https://www.artstation.com/blogs/luismesquita/PwEm/everything-about-pbr-textures-and-a-little-more-part-1
18. https://80.lv/articles/planning-the-key-to-environment-design/
19. "Egyptologist Answers Ancient Egypt Questions From Twitter." https://youtu.be/E7oEq6CE78g?si=IND5poD4XeTXZc53,3:51
20. https://docs.unrealengine.com/4.27/en-US/BuildingWorlds/Landscape/Creation/#:~:text=Before%20you%20can%20create%20a,Landscape%20tool%20at%20any%20time
21. https://docs.unrealengine.com/4.27/en-US/BuildingWorlds/Landscape/QuickStart/
22. https://docs.unrealengine.com/4.27/en-US/BuildingWorlds/Foliage/
23. https://www.thegnomonworkshop.com/tutorials/creating-foliage-for-videogames
24. https://docs.unrealengine.com/udk/Two/WorkflowAndModularity.html
25. Leszekpazderski: File: Luboszyce Palac 1 JPG—wiki. License: Creative Commons Attribution—Share Alike 4.0 international license (2014). File:Luboszyce palac 1.JPG—Wikimedia Commons
26. https://docs.unrealengine.com/4.27/en-US/BuildingWorlds/LightingAndShadows/
27. https://docs.unrealengine.com/4.27/en-US/Resources/Showcases/Effects/
28. https://docs.unrealengine.com/4.27/en-US/BuildingWorlds/FogEffects/
29. https://docs.unrealengine.com/4.27/en-US/WorkingWithAudio/SoundActors/

Chapter 6
Virtual Reality and Architectural Design for Industry

Yutaka Tada and Takashi Matsumoto

Abstract This chapter describes the importance of virtual reality (VR) in the field of architecture and the ways it benefits the industry in Japan. VR is a technology that enables architects to create immersive 3D environments, which is beneficial for the owner to understand the architect's intentions. Other applications include safety management in architecture construction and architectural education. VR is also effective not only in architecture, but also in consensus-making for urban projects. If the functional limitations of the Head Mounted Display are eliminated through technological development in the future, they can be used to a greater degree in the construction industry.

Keywords Virtual reality (VR) · Industry · Architectural design · VR use in the architectural design industry · VR benefits industry in Japan

6.1 Architectural Design Is a Black Box

Architectural design is a black box. The authors are currently working as researchers at a National Institute of Technology and have designed cities and architecture as professional engineers (Civil Engineers) and first class architects in local cities in Japan. Architectural design is complex, diverse, and individual. One of the reasons is that architecture is one of a kind. A building can only be built on a single piece of land on the earth. No two sites are the same in terms of latitude and longitude, climate, sunlight, frontage roads, surrounding architecture and parks, and the unique

Y. Tada (✉)
National Institute of Technology, Anan College, Anan, Japan
e-mail: y_tada@anan-nct.ac.jp

T. Matsumoto
National Institute of Technology, Tokyo, Japan
e-mail: matumoto@anan-nct.ac.jp

D. M. Barry, H. Kanematsu (eds.), *Applications of Metaverse and Virtual Reality to Creative Education and Industry*, Intelligent Systems Reference Library 267, https://doi.org/10.1007/978-981-96-3341-8_6

history and culture of each site. In a society with an industrial building dominated by agriculture, buildings are constructed using materials indigenous to the land, but in a society with an industrial building dominated by industry, steel-framed buildings, reinforced concrete buildings, and in recent years, cross-laminated timber buildings are used. In industrial societies with predominantly industrial buildings, steel-framed buildings and reinforced concrete buildings are used.

Another reason for the black box of architecture design is that each architect or client is different. First, a client tries to organize and communicate his criteria in as much detail as possible in order to accurately convey what he is paying for and what he wants. However, when 10 architects read and accurately understand the detailed criteria, every building produced will be different. The tens of thousands of architectural competitions that take place around the world every year attest to this. We can never fully communicate or understand everything in terms of others. The background and education of each architect is intricately intertwined in the creation of an architectural work. Similarly, the client has lived many different lives and has many different memories, feelings, and demands concerning the architecture. Unfortunately, it is not possible to completely verbalize all of them, nor is it possible to accurately convey them in words.

Nevertheless, we must discuss the architecture to be created with the client and the architect using various methods, such as language, drawings, models, perspectives, and sometimes poetry and images, until both parties are satisfied. The act of architecture is not an artistic activity, but a design to change society for the better. In Japanese, design means working with intention. In recent years, VR (virtual reality) and AR (augmented reality) have joined the ranks of new methods of expression to explain the architect's intentions. In this chapter, the authors discuss the value and potential of VR based on their experiences in architectural design and architectural education in local cities.

6.2 Possibility of VR to Represent Beauty in Architecture

The three rubrics of architecture as described by the Roman architect Vitruvius in his De architectura libridecem in the first century B.C. are Firmitas, Utilitas, and Venustas. Architecture cannot exist without any one of these rubrics. The harmonization of these three rubrics is the task of the architect. Firmitas is the easiest of these to explain objectively, and can be expressed in terms of physical laws, such as how many earthquakes or typhoons a building can resist based on structural calculations, or how much a floor will deflect when used for a long time. By applying such physical laws, it is possible to explain how many decades it will take before inconveniences occur in use and how much maintenance should be performed. In many countries and regions, the laws governing the construction of buildings have been established, and the failure of structural calculations can be explained in language by a judge.

Next, Utilitas can also be expressed in part in numbers. For example, a certain formula can be found for the number of people in a movie theater and the appropriate number of toilet cubicles, or the height of a kitchen table that is easy to use for different heights or with different disabilities can be determined statistically. On the other hand, there are some things that cannot be explained objectively. In many Japanese houses, baths are rarely located near bedrooms or other rooms with a high degree of privacy; on the contrary, they are often located near living rooms. In the democratization movement after World War II, architectural planners recommended a layout in which bedrooms and baths are located close to each other from the viewpoint of functionality, and even today, there is a national consciousness that baths should be located near family gathering places. There are a few exceptions in the Firmitas and Utilitas, but they can be explained using an objective common language such as physical laws and statistics between the client and the architect.

However, what about Venustas, a building that is beautiful to 99 people may not be so to one person. While it may be possible to statistically process whether a building is beautiful or not, that one person may be an important client, and a hundred years from now, that one person's opinion that a building is not beautiful may have historical significance. Therefore, the architect's effort to satisfy Venustas is also difficult to fully verbalize. It is somewhat well-known that people have certain feelings about architectural spaces. Architects design not the buildings themselves, but the margins that are enclosed or cut off by the buildings, i.e., the architectural space. For example, a sphere envelops us, giving us a sense of security and intimacy. A series of evenly spaced columns evokes a sense of order and rhythm, while curves in the golden ratio suggest harmony with the surrounding environment. However, while these can be described in a somewhat objective, common language, it is difficult to describe all of Venustas.

The most difficult way to explain Venustas is through drawings. What the architect ultimately wants to create is a three-dimensional architectural space. If one dimension is reduced from this, it can be expressed as a drawing. For example, a floor plan is a drawing of a building cut horizontally at eye level (1.5 m from the floor) and showing only the XY direction. By reducing the dimension in the Z direction (height direction), the building is seen from above. Figure 6.1b shows a 3D model of a house designed by a famous Japanese architect using Architecture Information Modeling (BIM). Students cut the model at 1.5 m from the height of the first floor to show a floor plan. It is difficult for ordinary people to understand the height direction just by looking at a floor plan, but architects have received such training so that they can see a three-dimensional architecture from a floor plan as shown in Fig. 6.1b. Similarly, a cross-sectional view is a drawing that only shows either the x- or y-direction and the z-direction (height direction), and it is expressed by eliminating depth. However, as shown in Fig. 6.1c, the architect can understand not only the height direction but also the depth direction from the drawing. In most cases, architects do not acquire the ability to explain the height direction in their daily lives, but learn it through making models (Fig. 6.2) and taking measurements of actual spaces (Fig. 6.3).

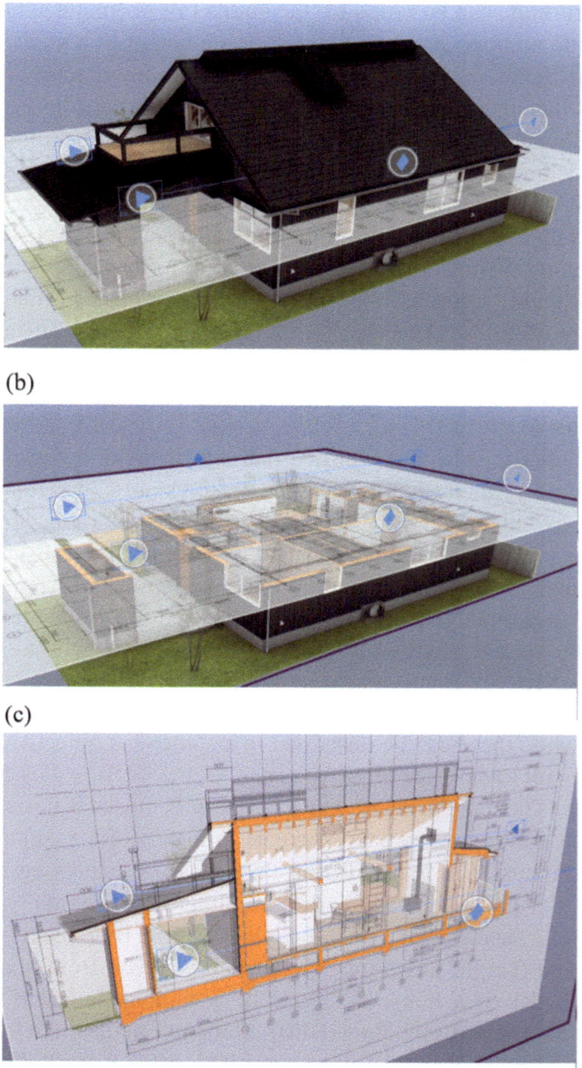

Fig. 6.1 3D models created by students (**a**) Panoramic view, (**b**) Plan view location representation, (**c**) Positional representation in cross-sectional view

Venustas in architectural space is expressed in the direction of height, i.e., in cross sections. One of the most famous cross sections in the world (Fig. 6.4a) is the Pantheon in the city of Rome. The present Pantheon was rebuilt by the Roman Emperor Hadrian between 118 and 128. It consists of a cylindrical and hemispherical dome on top of a circular dome 43.2 m in diameter, which is 43.2 m high as the diameter. At the top of this dome is a round skylight (Fig. 6.4b), through which light shines from above. The circle on which the light is projected moves moment by

Fig. 6.2 Architectural models created by students

Fig. 6.3 Students taking measurements

Fig. 6.4 The Pantheon; (**a**) Section [1], (**b**) Inside view

(a)

(b)

moment on the floor and walls according to the movement of the sun. Voices from the surrounding plaza also pour down like thunderstorms.

The scene evokes a magnificent universe and the great order that governs it. In order to be able to create such architectural spaces from cross-sectional drawings, it is considered important to visit architectural works and experience the space. In recent years, research has been conducted to see if the spatial experience in VR has the same effect as visiting actual architectural works. For example, Ishida et al. (2019) [2] captures the development of spatial cognition even among those who do not specialize in architectural design through the experience of VR spaces. Hou et al. (2021) [3] and Hou et al. (2022) [4] show that, through VR, students can learn a sense of scale not only in indoor spaces but also in larger-scale outdoor spaces. In addition, while overseas travel was banned for a while because of the COVID-19 pandemic, VR tours have been launched at famous sightseeing spots such as the Pantheon, creating an environment that facilitates spatial experiences. In any case, the ability to explain Venustas is acquired through training to verbalize the reasons why one feels the architectural space is beautiful, whether in real space or in VR space.

The first case in which the author utilized VR to explain Venustas was the renovation of a hall in a children's science museum in a local city in 2017. The hall was a mortar-shaped cylinder with a 10-m radius and a 10-m height to the ceiling, where a Foucault pendulum was suspended from the ceiling and displayed, but there was no spatial unity. The author proposed to install a rose-curved membrane on the ceiling surface, which is expressed by the movement of the pendulum, and to explain the exhibition in unison with Foucault's pendulum. Figure 6.5a, b are drawings of the proposal. However, it was difficult to accurately convey the architectural space proposed by the author to the architectural staff of the local government, who placed the order, only with these drawings. Therefore, a 3D model was created as shown in Fig. 6.5c, and the architects responsible for the local government were given the opportunity to experience the space in VR. This allowed them to better understand the author's intent - namely, to create a sense of unity within the space through the use of rose curves and to evoke warmth by designing each individual membrane with a swelling form. In the construction phase, we provided VR data to the builders so that they could understand the architect's intentions and work on the project. The final completed space is shown in Fig. 6.5d, and we are proud of the high accuracy of the VR representation. It is important to note that architects do not use VR as a reference to create architectural spaces, but rather use 3D models and VR to convey the architect's image of the architectural space in a way that is easy to understand. In particular, the current development of BIM makes it easy to create beautiful perspectives and VR. If this were reversed, the architect's creativity would be lost, and 3D models and what can be expressed in 3D would be all there is to convey expression.

6.3 Potential of VR in Architecture Industry

In the previous section, we discussed the possibilities and limitations of VR as a method of expressing architectural space, especially from the aspect of architectural beauty. In this section, we describe the possibilities for the entire architectural industry, including function and building.

(a)

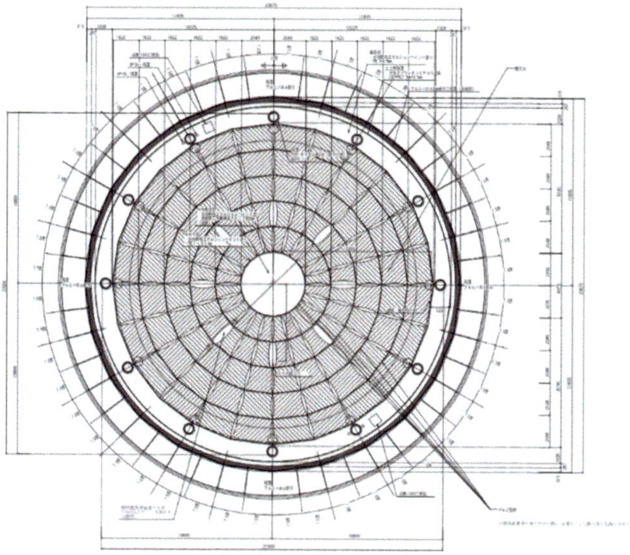

(b)

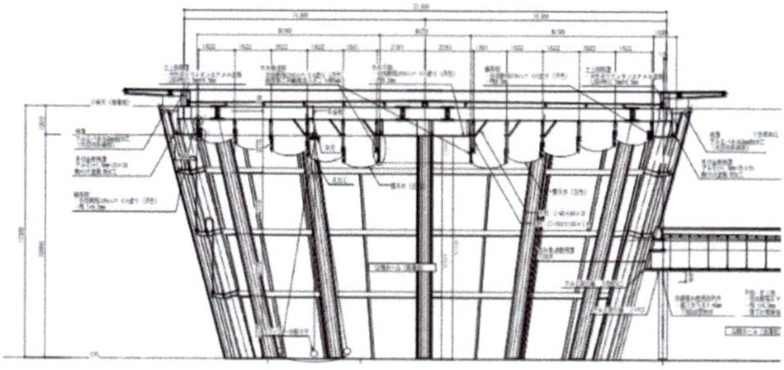

(c)

Fig. 6.5 Children's Science Museum Hall; (**a**) Ceiling Plan, (**b**) Section, (**c**) Design model, (**d**) Completed works

(d)

Fig. 6.5 (continued)

First, the housing sector accounts for a large portion of the architecture industry in Japan. After World War II, Japan adopted a policy of owner-occupied housing, and by 1973, when official statistics were available, the percentage of owner-occupied housing was approximately 60%, and has remained at around 60% to this day. More detached houses are newly built in rural areas than in urban areas. According to the Ministry of Land, Infra-building, Transport and Tourism (2023) [5], about 50% of the households that built new detached houses obtained information on housing construction companies at housing exhibitions. In Japan, there are few opportunities to learn about architecture in compulsory education, and few members of the general public have sufficient personal opinions when designing their own houses. For this reason, many construction companies specializing in housing have built model houses to allow builders to experience the space of their housing products. The use of VR in housing exhibition halls began with advanced efforts by ONO (2005) [6] and others, and since the Covid-19 pandemic, national and local housing specialty construction companies have opened VR exhibition halls.

However, according to MLIT (2023), only 5.7% of households that have newly built detached houses have previewed properties using VR or AR tools. According to Ishimura (2021) [7], the awareness rate of VR in Japan is over 90%, but the number of people using VR on a daily basis is about 5%. In addition, the personal ownership rate of HMDs (head mounted displays) among Japanese is only 6.0%. Reasons for the lack of daily use of VR include the high cost of HMDs, the lack of progress in weight reduction, and the feeling of fatigue when wearing HMDs. In particular, with regard to fatigue, according to Oyama (2002) [8], the maximum time for using VR in a housing exhibition hall is about 12 min, and such a short time

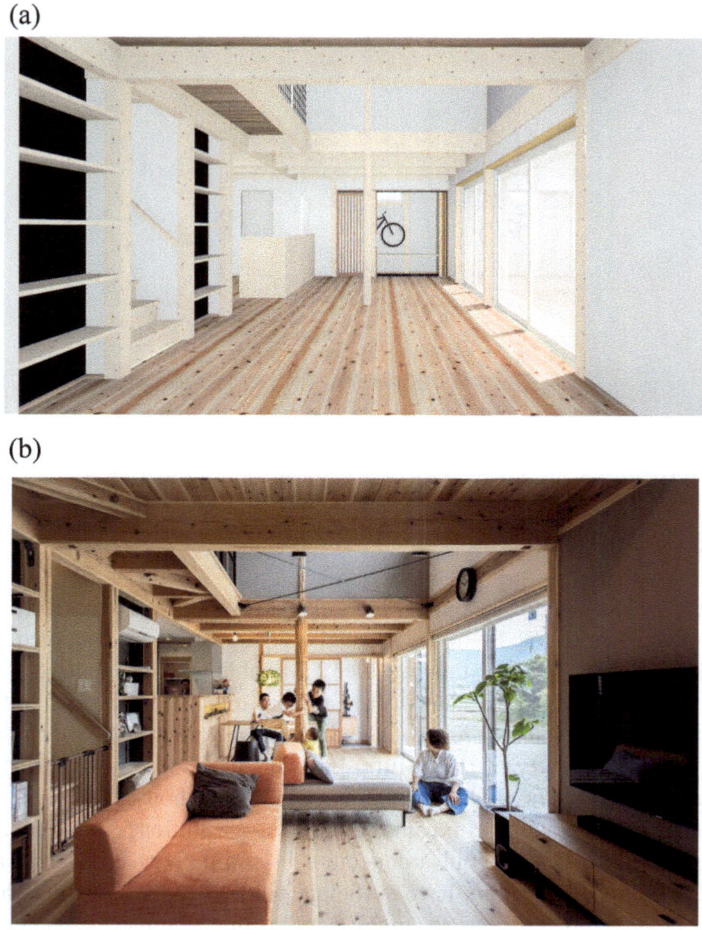

Fig. 6.6 Example of a wooden house designed using VR (**a**) Design model, (**b**) Completed works

may serve as an accent during meetings, but it cannot become the mainstream way to show off designs.

The author also used VR in house design, and utilized it for partial meetings such as checking the connection between rooms, checking the direction of flooring and ceiling boards (Fig. 6.6a, b), and understanding how the windows of the house next door look from his room. At present, many people are not accustomed to wearing HMDs, and it is not yet realistic to have them learn how to use HMDs and conduct all meetings using HMDs in the limited time available for meetings with clients whose parents both work and have small children. Improvement of the HMD itself is required for its widespread use in the construction industry.

Currently, the number of new houses in Japan continues to decline due to a falling birthrate and an aging population. The annual number of new housing starts dropped below 1 million in 2009 and further declined to 860,000 by 2022. This downward trend is expected to continue in the future [9] and the number of new houses is expected to fall to 550,000 units by 2040. In contrast, the remodeling market is expected to remain steady at around 1 trillion yen, and demand for inspections to evaluate the deteriorated condition and performance of houses before remodeling is increases in Japan. However, Tada (2022) [10] points out that the quality of inspections by inspectors varies, and a VR system for training inspectors is currently under development. VR is highly effective for such educational purposes. For example, construction companies use VR for safety education to prevent mistakes. In our school, when students construct a wooden booth, they not only check the 3D model as shown in Fig. 6.7a, but also experience the assembly procedure by VR as shown in Fig. 6.7b, to enhance safety.

(a)
(b)

Fig. 6.7 Students check the process during the construction of a wooden booth; (**a**) Overall explanation is using 3D models. (**b**) Individual explanations using VR

Fig. 6.8 Office workers practicing VR as recurrent education

Our school provides recurrent education for engineers and office workers who work for local construction companies specializing in housing to create and experience VR as shown in Fig. 6.8. The reason why we chose office workers as the target of our recurrent education program is that in order for VR to be used as a matter of course in the construction industry, not only engineers but also everyone, including office workers, should be able to use it. All participants will be able to insert their own BIM models into VR soon.

6.4 Potential of VR in Urban Planning

In the field of urban planning, VR is increasingly being used to obtain the consent of residents for urban development projects. Since 2021, the authors, in collaboration with the University of Tokyo, the University of Tokushima, and others, have been conducting workshops with local residents and government officials in Anan City, Tokushima Prefecture, to formulate a preliminary reconstruction plan. The Nankai Trough earthquake, which has a probability of occurrence exceeding 70% within the next 30 years, is expected to cause earthquake and tsunami damage over a wide area from the suburbs of Tokyo to eastern Kyushu, resulting in 320,000 fatalities and more than one million buildings destroyed. Anan city has a coastal railroad line, and residential and commercial buildings are concentrated around the railway stations. The coastal area has soft strata and high surface ground amplification, which causes large shaking during earthquakes, and the number of destroyed buildings reaches about 16,000 (about 60% of all buildings in the city) [11]. The

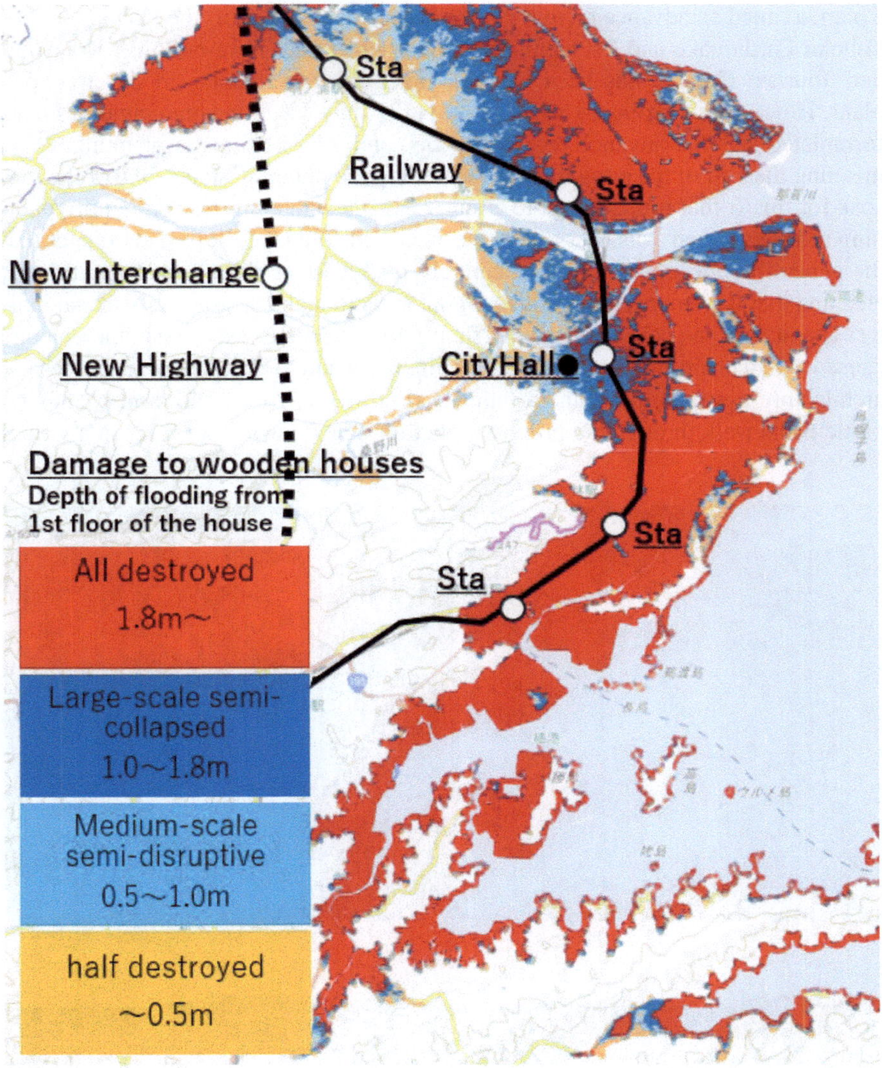

Fig. 6.9 Predicted damage to housing in the event of the largest possible tsunami (Nankai Trough Earthquake)

tsunami is expected to reach up to 3 km from the coast, with a maximum tsunami depth of 10 m. Figure 6.9 shows the estimated inundation depths, color-coded according to the buildings damage criteria of the Cabinet Office (2021) [12], which indicates that wooden houses will be totally destroyed in most coastal areas. The number of fatalities is estimated to be 4600, or 6.1% of the total population.

To prevent such damage, the authors propose that the city be relocated around the Anan Interchange of the newly planned high-standard road in the inland area. Such a plan is called an advance recovery plan, and since the 2011 off the Pacific coast of Tohoku Earthquake and Tsunami, the Ministry of Land, Infra-building, Transport and Tourism (MLIT) has been encouraging local governments to develop such plans. However, according to MLIT (2023) [13], only 7% of the municipalities have formulated specific advance recovery plans, and 17% are in the planning stage, meaning that 3/4 of the municipalities are not in the planning stage. One of the reasons for this is that it is difficult to assume damage and consider community planning at a stage when a disaster has not yet occurred. Tada (2023) [14] also states that the general public is not able to accurately read disaster risks from hazard maps, etc., and that it is important to utilize digital tools to make disasters a personal matter. The author has developed a workshop with citizens to experience the damage caused by a tsunami in Anan City in 3D (Fig. 6.10) and to experience the process of architecture destruction by an earthquake (Fig. 6.11a, b). In addition, by having participants walk through the proposed plan for a new town (Fig. 6.12) in a virtual

Fig. 6.10 Three-dimensional data on expected tsunami damage in Anan City

(a)

(b)

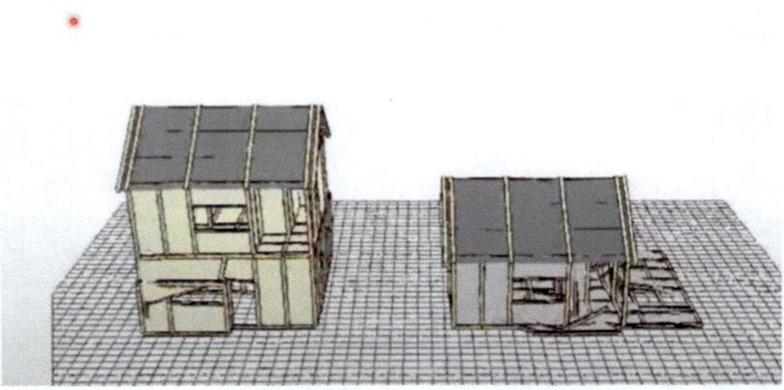

Fig. 6.11 Simulation of earthquake-induced collapse of wooden buildings. (**a**) Earthquake-induced collapse of a wooden building at E-Difference, (**b**) Earthquake-induced collapse of a wooden house using the collapse simulation "wallstat" [15]

reality using VR, etc., we encourage the enjoyment of creating a new town and the feeling of support for it.

In Japan, point cloud data for each city is beginning to be accumulated at PLATEAU [16]. By utilizing this data, it will be easy to simulate how a city changes with the development of roads, railroads, architectures, parks, etc. using VR. The accumulation of such urban data that can be used free of charge is a necessary condition for utilizing VR.

Fig. 6.12 New urban vision around the new interchange

6.5 What Is Required for the Use of VR in the Construction Industry

In this chapter, Sect. 6.1 describes that architectural design is a black box, and Sect. 6.2 shows that VR is particularly effective in explaining the designer's intention regarding beauty in architecture and in educating architects. Section 6.3 focuses on the housing industry, and shows that VR is currently used only as a supplementary role in design due to functional limitations of HMDs, and that VR is considered effective for safety management during construction. Section 6.4 expands the viewpoint to the urban planning field, and shows that VR is effective in consensus architecture for urban projects. Finally, in this section, we present our personal views on the technological development required for the further utilization of VR in the construction industry.

Over the past few decades, 3D CAD, Adobe Illustrator, and Photoshop have become popular, and students with little experience are now able to create designs that previously could only be drawn by professionals. Previously, students who could not draw could not play an active role in the architectural industry, but this is no longer the case. With the spread of 3D CAD and BIM, students who could not immediately visualize 3D space from 2D drawings were able to comprehend 3D space and play an active role in the architectural industry. Nowadays, AI can automatically perform a part of architectural design. In this way, technological advances are expanding the range of people who can be involved in architecture. What kind of people will VR attract to the field of architecture?

First, architects capable of intense three-dimensional modeling can expand the possibilities for their talents. Architecture is a design, not an art, and nothing progresses without the client's understanding, except in the unbuilt world. First-of-its-kind architectural spaces will continue to be created, but they will require a client who understands the design, and it is imagined that the experience of simulating such a space will deepen this understanding. At the same time, this will expand the possibilities for the client to obtain a new architectural space.

In the future, what will happen if VR is augmented with not only visual information but also auditory information such as the sound of wind and water, tactile information such as the sensation of walking barefoot on the floor and the temperature, and olfactory information such as the aroma of tatami mats and cedar wood? If these were realized, we would be able to experience the smell and taste of warm coffee in a café that has been renovated to match the traditional townscape, while listening to music and the casual conversation of the people. If such technology is realized, it will be possible to explain the architect's intentions in a dimension completely different from that of conventional methods of expression, drawings, perspectives, and other visual information. Whether the black box will become smaller or not remains to be seen, but the number of stores that serve the black coffee of your choice may increase.

References

1. Banister Fletcher: A History of Architecture (1975)
2. Ishida, K., Sakatani, S., Tanaka, Y., Chiba, M.: Influence of experience of space using VR on the design process. J. Architect. Plan. (Trans. AIJ). **84**(761), 1579–1587 (2019)
3. Hou, N., Nishina, D., Sugita, S., Jiang, R., Oishi, H., Kindaichi, S., Shimizu, A.: A study on possibility of using vr space in design education part 1: verification of vr space effectiveness by learning experiment of scale feeling. J, Environ, Eng, (Trans, AIJ). **86**(785), 670–679 (2021)

4. Hou, N., Nishina, D., Sugita, S., Jiang, R., Kindaichi, S., Oishi, H.: A study on possibility of vr space in design education. AIJ J. Technol. Des. **28**(70), 1254–1259 (2022)
5. Ministry of Land, Infrastructure, Transport and Tourism, Housing Bureau, Residential Market Trend Survey Report for FY2022 (2023). https://www.mlit.go.jp/report/press/content/001610299.pdf
6. Ono, K., Ozato, M., Yuyama, C., Tokunaga, Y., Hashiguchi, H., Hirate, K.: Development of virtual simulation system for housing environment using mixed reality technology (Information Systems Technology). AIJ J. Technol. Des. **11**(22), 567–572 (2005)
7. Ishimura, N., Mamiya, D., Kato, Y.: Challenges for Japanese companies in the platform competition for AR/VR. Dev. Bank Jpn. Res. **354**(1) (2021)
8. Oyama, Y., Ono, K., Morikawa, Y., Yoshizawa, N.: Research on the presentation and evaluation of living spaces using VR. TAISEI Techn. Centre Bull. **35**, 38-1–38-6 (2002)
9. Nomura Research Institute Group, New housing starts fall to 550,000 in 2040
10. Yutaka, T.A.D.A.: A study on criteria for performance improvement inspection for wooden detached houses, part 2. Private businesses. AIJ J. Technol. Des. **28**(69), 762–767 (2022)
11. Tokushima Prefecture Government, Publication of Tokushima Prefecture Nankai Trough Large Earthquake Damage Assumption (Second Assumption) (2013). https://anshin.pref.tokushima.jp/docs/2013112100023/
12. Cabinet Office: Operational guidelines on the criteria for recognition of damage to dwellings resulting from disasters, Part 2: damage caused by flooding (2021). https://www.bousai.go.jp/taisaku/unyou.html
13. Ministry of Land, Infrastructure, Transport and Tourism: Guidelines for the Examination of Preliminary Reconstruction Town Development Plans (2023)
14. Tada, Y., Kato, K., Shiozaki, Y., Suzuki, S.: New ways to utilize flood hazard maps to select appropriate residential area. J. Jpn. Soc. Civil Eng. F6 (Saf. Prob.). **78**(2), I_105–I_122 (2023)
15. Nakagawa, T., Ohta, M., et al.: Collapsing process simulations of timber structures under dynamic loading III: numerical simulations of the real size wooden houses. J. Wood Sci. **56**(4), 284–292 (2010)
16. https://www.mlit.go.jp/plateau/

Chapter 7
Application of Virtual Reality to Medical Education

Takahisa Kamikura and Hideo Inaba

Abstract Virtual reality (VR) has been increasingly used in medical education. VR simulations can provide learners with a safe and immersive environment to practice their skills and knowledge. These simulations can range from emergency situations to those for exploring human anatomy. This chapter provides examples of VR use that focuses on emergencies and critical care. VR education for these situations provides invaluable medical simulations from resuscitation to managing life threatening cases with hands-on experiences without real life consequences. It allows trainees to develop critical thinking, decision-making skills, and teamwork under realistic conditions. VR can also be used to teach various skills for surgical procedures, cardiopulmonary resuscitation (CPR), and other important skills.

Keywords Virtual reality (VR) · Medical education · VR simulations for medical training

7.1 Introduction: Overview of Virtual Reality in Healthcare and Medicine

Virtual Reality (VR) is a computer-generated, simulated 3D environment [1]. It can provide users with an immersive experience that feels like reality to varying degrees, ranging from a simple presentation on a 2D display screen system (desktop VR) to highly immersive systems (immersive VR) that use head-mounted displays (HMDs) [1–4] The breakthrough advancements in this technology have revolutionized the

T. Kamikura (✉)
Suzuka University of Medical Science, Suzuka, Japan
e-mail: kamikura@suzuka-u.ac.jp

H. Inaba
The Department of Emergency Medicine, Kanazawa Medical University, Uchinada, Japan
e-mail: hidinaba@kanazawa-med.ac.jp

D. M. Barry, H. Kanematsu (eds.), *Applications of Metaverse and Virtual Reality to Creative Education and Industry*, Intelligent Systems Reference Library 267, https://doi.org/10.1007/978-981-96-3341-8_7

medical field. We should not miss out on the opportunity to witness the positive impact that VR can have in the medical field.

In recent years, VR technology has brought about a significant transformation in the healthcare field. It has proven to be an effective tool for treating a variety of health conditions, as well as for medical education and training [5–10]. VR is being used in physical rehabilitation to improve patients' motor function, fitness, movement quality, and mobility [11–14]. Patients can engage in interactive exercises and simulations that mimic real-world scenarios, allowing them to practice movements and activities that challenge their abilities. This can lead to more effective and efficient rehabilitation outcomes.

Moreover, VR has particularly shown success in addressing mental disorders such as anxiety and depression [15–22]. By providing exposure therapy in a safe and controlled environment, patients are able to confront their fears.

Additionally, VR has proven to be an enjoyable method for managing pain [23, 24]. It provides a distraction from the pain by immersing patients in an engaging virtual environment. This can reduce the need for medication and provide a non-invasive solution for pain management [25].

Regarding education and training, VR was greatly appreciated by medical and nursing students [2, 26–30], and proved to play a crucial role in improving medical knowledge [31–34] and fostering clinical skills [35–37].

Overall, VR technology has opened up a world of possibilities for the medical and healthcare field, providing innovative solutions to long-standing challenges. In this chapter, we provided examples of VR use that focuses on emergency and critical care.

7.2 Mass Casualty Triage Training

Mass casualty incidents triage systems are implemented to offer the greatest good to the greatest amount of people as healthcare resources are limited or strained due to the number of injured individuals [38]. Since the early control trials reported by Vincent DS et al. [39] and Andreatta RB et al. [40], several studies [41–44] reported the efficacy of VR in mass casualty triage training (Table 7.1). A recent systematic review showed that high-fidelity VR training should not rule out live simulation but rather serve as a complement [45]. The use of high-tech simulations like VR and artificial intelligence (AI) has greatly increased during and after the COVID-19 pandemic, resulting in positive learning outcomes. There is a concern about using VR that relates to the technical problems that users find distracting and frustrating. Moreover, some first responders have reported experiencing motion sickness and dizziness while using the VR system [3, 45], which could pose a challenge to its implementation. Therefore, efforts should be made to address these technical issues and reduce usability problems.

Table 7.1 Virtual reality in triage training

Study	Partcipants	Triage system	VR intervention	Control	Results
Vincent DS et.al (2008) [38]	24 medical students	Mass casualty triage (MCI)	Head-mounted display (HMD) and three motion tracking sensors	None (Time series analyses)	Improved triage and intervention scores, speed, and self-efficacy during an iterative, fully immersed VR triage experience
Andreatta PB et.al (2010) [39]	15 postgraduate residents with postgraduate years of 1–4	Simple triage and rapid treatment (START)	a full-immersion VR environment enclosed by four walls, floor, and ceiling to create a facsimile of "reality" using sophisticated three-dimensional computer-based imaging and interactivity (CAVE)	Standardized patient (SP)	No significant differences between the triage performances of the VR and SP groups, but the data showed an effect in favor of the SP group performance on the posttest
Luigi Ingrassia P (2015) [40]	46 students in their last year of medical school	Simple triage and rapid treatment (START)	VR scenario	Live scenario	Virtual reality simulation proved to be a valuable tool, equivalent to live simulation
Lowe J et al. (2020) [41]	207 subjects at the American College of Emergency Physicians (ACEP) 50th annual conference in San Diego, California from October 1–3, 2018 (hours/ day). Subjects	Pesiatric mass casualty incident (MCI)	A 360 VR MCI module	None	Participants felt the 360 VR experience was engaging (median = 5) and enjoyable (median = 5). Most felt that 360 VR was more immersive than mannequin-based simulation training (median = 5)

(continued)

Table 7.1 (continued)

Study	Partcipants	Triage system	VR intervention	Control	Results
Mills B et al. (2020) [42]	29 second-year paramedicine students	Mass-casualty incidents (MCIs)	A bespoke virtual-reality (VR) MCI simulation	Live simulation scenario	VR simulation provided near identical simulation efficacy for paramedicine students compared to the live simulation
Harada et al (2023) [43]	70 paramedic students classified into VR (N = 33) and live lecture (N = 29) groups	Simple modified START methods	A cloud VR system	Live lecture	The test score was higher in the VR group

7.3 Basic Life Support (Basic Cardiac Life Support/Cardiopulmonary Resuscitation)

Basic Life Support (BLS) comprises a set of life-saving procedures that utilize cardiopulmonary resuscitation (CPR) to help maintain a patient's circulation and respiration until advanced life support arrives. This involves properly identifying the occurrence of cardiac arrest, calling for emergency medical assistance, providing early and effective CPR with minimal disruption to chest compressions, and promptly utilizing an automated external defibrillator (AED). Early and accurate BLS intervention helps improve oxygenation and increases the likelihood of survival. CPR is administered to individuals experiencing cardiac arrest. The quality of CPR, specifically the rate and depth of chest compressions performed by bystanders who witness cardiac arrest, is crucial in determining favorable neurological outcomes of out-of-hospital cardiac arrest (OHCA) [46–48].

There are two types of CPR: hands-only CPR (also known as compression-only CPR) and conventional CPR (also known as traditional CPR) with rescue breaths. If bystanders have not received CPR training before, received training long ago, or lack confidence even after being trained, it is recommended that they use hands-only CPR. Only bystanders who have received training in conventional CPR and are comfortable with the technique can use it [49].

BLS training is crucial in reducing sudden cardiac arrest deaths. Traditional instructor-led CPR training has limitations, including the need for retraining, specific spaces and schedules, and the lack of realism and immersion of traditional feedback devices [50]. The COVID-19 pandemic reduced the availability of effective instructor-led CPR training [51].

Virtual Reality (VR) can be an effective tool for teaching cardiopulmonary resuscitation (CPR) to a larger number of people, improving their knowledge and skills, and providing more convenient and frequent learning opportunities. The American Heart Association supported the use of VR in CPR education in 2018, recognizing the potential of immersive technologies and gamified learning to enhance the learning experience [52]. An evidence update from ILCOR CoSTR [53] suggested that simulation-based resuscitation education can be integrated into continuous education programs for life support courses, either through in situ training or in a dedicated simulation center. The 2021 European Resuscitation Council Guidelines recommended incorporating virtual learning environments into all levels of CPR training as part of a blended and self-learning approach [54].

Virtual reality has some limitations when it comes to teaching BLS. The primary issue is that certain actions, such as chest compressions, ventilation, and defibrillation, cannot be performed with the same level of accuracy as they would in a real-life setting. This is because VR relies on haptic controls to carry out these actions and can't provide the same level of physical feedback as real-life scenarios.

To overcome these limitations, extended reality and/or augmented reality can be integrated into the training [55–62]. This can include full-scale mannequins and simulation objects, such as AEDs. By adding these elements, VR can be used to simulate a real cardiac arrest scenario, providing users with a more realistic learning experience (Table 7.2). This helps them to learn how to identify cardiac arrest and perform the necessary steps in the chain of survival until emergency medical services arrive.

Digital resuscitation training, which includes blended online games, computer support, and mobile or virtual learning, was introduced in the 2010s. This was before advanced VR training became common. Lau Y et al. [63] conducted a systematic review and evaluation of the overall efficacy of digital resuscitation training. The review included 20 randomized control trials (RCTs) [64–83]. The meta-analyses performed in this review evaluated three scores: knowledge, skill performance, and correct compression rate.

The review found that the digital resuscitation training significantly improved knowledge. However, the improvement was further enhanced by the application of learning theories and video-recorded assessments among health professionals. The overall skill performance scores were not considerably improved by digital resuscitation training. However, when trainees used computer screens through the video-recorded evaluation, their skill performance was significantly improved. On the other hand, digital resuscitation training significantly reduced the correct compression rates compared to standard resuscitation training. This was particularly noticeable when trainees used online training with a forum.

A recent review of studies on Virtual Reality (VR) in Basic Life Support (BLS) training has concluded that it can help improve the manual skills and self-efficacy of adult non-healthcare laypersons [84], The review included four randomized controlled trials [85–88] and two quasi-experimental studies [55, 89] conducted between 2017 and 2022. However, the studies were of low quality, and it is uncertain if VR can lead to better outcomes for patients experiencing cardiac arrest in community settings. Future studies should aim to address this gap by obtaining long-term outcomes and looking at other factors such as survival rates, return to

Table 7.2 Advanced VR system employed in studies reported in 2020s

Authors, year	Country	System	Main findings
Jaskiewicz F, Kowalewski D, Starosta K, Cierniak M, Timler D. 2020 [56]	Poland	VR prototype (beta version of CPR virtual reality learning software—VR ACT, Octopus VR, Lodz, Poland) HTC Vive (HTC, Taoyuan, Taiwan) Noitom Hi5 VR Glove (Noitom International Inc. Miami, Florida, USA) CPR mannequin integrated in the virtual space and covered with a virtual 3D-human model	97.8% of respondents believe that training with the use of VR is more effective than a traditional method ($P < .01$) Additional VR equipment placed on the trainee's body may cause chest compressions harder to provide
Issleib M, Kromer A, Pinnschmidt HO, Süss-Havemann C, Kubitz JC. 2021 [57] Moll-Khosrawi P, Falb A, Pinnschmidt H, Zöllner C, Issleib M. 2022 [58]	Germany	The VR software developed in cooperation between Universitätsklinikum Hamburg Eppendorf and VIREED The Leardal® QCPR Mannequin connected to the VR system A visual feedback on the quality of chest compressions. Bag-mask-ventilation and the use of an AED are virtually implemented in the system (but no actual haptic handling takes place)	A higher learning gain in 6 out of 11 items of the questionnaire A significant lower no-flow time A better BLS performance
García Fierros FJ, Moreno Escobar JJ, Sepúlveda Cervantes G, Morales Matamoros O, Tejeida Padilla R. 2021 [59]	Mexico	Virtual^{CPR} system composed of the HX711 integrated circuit block, an analog–digital convet, the sen-1045 sensor Serial communication with the NODEMCU ESP32 microcontroller A program is designed for the microcontroller that receives the serial data from the converter and, using its transfer function, obtains the equivalence in kilograms	The more previous training in CPR a user of the Virtual^{CPR} system has, the greater the percentage of correct compressions obtained from a virtual CPR session

(continued)

Table 7.2 (continued)

Authors, year	Country	System	Main findings
Sadeghi AH, Peek JJ, Max SA, Smit LL, Martina BG, Rosalia RA, Bakhuis W, Bogers AJ, Mahtab EA. 2022 [60]	Netherlands Republic of North Macedonia	Unreal Engine (Epic Games, Cary, North Carolina) software An Oculus Quest 2 (Oculus, Irvine, California) head-mounted display (HMD), in combination with two VR controllers and a high-performance laptop	13 (87%) of the expert participants would recommend VR training to other colleagues, and 14 (93%) of the expert participants thought the CPVR-sim was a useful method to train for infrequent post–cardiac surgery emergencies requiring CPR
Kim EA, Cho KJ. 2023 [61]	Korea	Medi-VR Simulation for Prehospital Cardiac Arrest Emergency Care Flipped Learning for Prehospital Cardiac Arrest Emergency Care as control	The post-education scores for CPR performance knowledge and CPR performance were significantly higher in the medi-VR simulation group compared to the flipped learning counterparts
Lee DK, Choi H, Jheon S, Jo YH, Im CW, Il SY. 2022 [62]	Korea	Hand-tracking technology (Leap Motion tracker) VR manekins (Nurugo B100, BEST CPR Inc., Gimpo, South Korea) equipped with five sensors 3DS Max (Autodesk, San Rafael, USA) Graphical User Interface (GUI)	The system can conduct BLS education without requiring instructors and trainees to gather

spontaneous circulation, and neurologically favorable outcomes. VR may be a useful teaching method in a blended learning BLS training strategy to improve the outcomes of out-of-hospital cardiac arrest in the community.

7.4 Advanced Life Support (Advanced CPR)

Advanced Life Support (ALS) refers to a range of life-saving techniques and skills that medical professionals use to assist in circulation, breathing, and ventilation beyond basic life support (BLS) [90]. ALS may involve procedures such as initiating intravenous access, reading and interpreting electrocardiograms, and administering emergency medications. All medical staff with critical care exposure are required to undergo ALS training as part of the ALS team. Alongside technical skills, non-technical skills such as communication, teamwork, information sharing, and situational awareness are also crucial components of ALS training [52, 91, 92].

Currently, ALS training in many hospitals includes pre-reading from a manual outlining the procedure algorithm, quizzes with multiple-choice questions, face-to-face training days including skills stations, didactic sessions, simulations, and a face-to-face assessment and qualification. The initial face-to-face training, refresher training, and reaccreditation processes are resource-intensive and require high instructor-to-participant ratios and costly simulation equipment.

A recent RCT [93] compared a VR-based serious gaming module called "3DMedsim ALS VR" with conventional training for ALS. The serious game module included a 3D visualization engine integrated with a learning record score that tracks users' actions and generates experience application programming interface calls for each action. The RCT showed that the majority of participants preferred the VR-based ALS serious gaming module over conventional lecture-based training. However, overall performances in the manikin-based simulation session did not significantly differ between the two training groups. Further large-scale studies are needed to determine the learning outcomes of VR-based ALS training.

7.5 Advanced Trauma Management (Advanced Trauma Life Support)

Healthcare systems are confronted with a major challenge in treating major trauma, which still remains the primary cause of death among the younger population. However, the outcomes of these patients have been significantly improved through various educational and systemic enhancements [94, 95].

Despite the considerable benefit of VR in medical education, few articles on the application of VR in advanced trauma life support (ATLS) education have been published, presumably because the major organizations manage official ATLS-related courses around the world [96–98]. One projective study reported that VR teaching improved the doctor's performance at managing major trauma patients and decreased doctor anxiety towards management and exposure to these clinical situations [99]. In a recent article [100], authors recommended preparing VR techniques combined with artificial intelligence (AI) and ultra-fast communication between users (5 G).

7.6 VR in Other Technical Components of Emergency Medicine

In education and training for emergency medicine, it is crucial to identify the best ways to use VS technology. As mentioned in the section on ALS, interprofessional communication and information sharing are critical for healthcare teams. Multiuser VR simulations or single-user VR with artificial intelligence (AI) help multidisciplinary teams practice, rehearse, and debrief together [101, 102]. Virtual reality is useful for training in infrequent high-acuity situations, like response to disasters or mass casualty events involving chemical [103], biological, nuclear, or explosive agents [104–106]. Virtual simulations can replace traditional simulations and provide more frequent practice. Additionally, virtual reality can assess rare clinical events or procedures [107, 108].

VR-based simulations are effective for assessing procedural skills [109–113], but realistic tactile sensation, or haptic feedback, is required to enhance the effectiveness of VR [114, 115]. We need to determine which areas of EM training would benefit from haptics training, 3D simulation, AR, and other visualization technologies.

7.7 Perspective and Conclusions

This chapter provided evidence for the efficacy of VR use that focuses on emergency and critical care. VR education for these situations provides invaluable medical simulations, from resuscitation to managing life-threatening cases with hands-on experiences without real-life consequences. To augment the efficacy of VR in this field, proper application of newly developed technologies and modification of VR-based education based on in backgrounds and characteristics of learners would be necessary.

References

1. Abbas, J.R., O'Connor, A., Ganapathy, E., Isba, R., Payton, A., McGrath, B., Tollet, N., Bruce, I.A.: What is virtual reality? A healthcare-focused systematic review of definitions. Health Policy Technol. **2**, 100741 (2023)
2. Barteit, S., Lanfermann, L., Bärnighausen, T., Neuhann, F., Beiersmann, C.: Augmented, mixed, and virtual reality-based head-mounted devices for medical education: systematic review. JMIR Serious Games. **9**(3), e29080 (2021). https://doi.org/10.2196/29080
3. Abbas, J.R., Chu, M.M.H., Jeyarajah, C., Isba, R., Payton, A., McGrath, B., Tolley, N., Bruce, I.: Virtual reality in simulation-based emergency skills training: a systematic review with a narrative synthesis. Resusc. Plus. **16**, 100484 (2023)
4. Matamala-Gomez, M., Bottiroli, S., Realdon, O., Riva, G., Galvagni, L., Platz, T., Sandrini, G., De Icco, R., Tassorelli, C.: Telemedicine and virtual reality at time of COVID-19 pandemic: an overview for future perspectives in neurorehabilitation. Front. Neurol. **12**, 646902 (2021)
5. Riener, R., Harders, M.: Virtual Reality in Medicine. Springer, London (2012)
6. Izard, S.G., Juanes, J.A., García Peñalvo, F.J., Estella, J.M.G., Ledesma, M.J.S., Ruisoto, P.: Virtual reality as an educational and training tool for medicine. J. Med. Syst. **42**(3), 50 (2018)
7. Liaw, S.Y., Carpio, G.A.C., Lau, Y., Tan, S.C., Lim, W.S., Goh, P.S.: Multiuser virtual worlds in healthcare education: a systematic review. Nurse Educ. Today. **65**, 136–149 (2018)
8. Rebelo, F., Noriega, P., Duarte, E., Soares, M.: Using virtual reality to assess user experience. Hum. Factors. **54**(6), 964–982 (2012)

9. Singh, R.P., Javaid, M., Kataria, R., Tyagi, M., Haleem, A., Suman, R.: Significant applications of virtual reality for COVID-19 pandemic. Diabetes Metab. Syndr. **14**(4), 661–664 (2020)
10. Li, L., Yu, F., Shi, D., Shi, J., Tian, Z., Yang, J., Wang, X., Jiang, Q.: Application of virtual reality technology in clinical medicine. Am. J. Transl. Res. **9**(9), 3867–3880 (2017)
11. Chen, J., Or, C.K., Chen, T.: Effectiveness of using virtual reality-supported exercise therapy for upper extremity motor rehabilitation in patients with stroke: systematic review and meta-analysis of randomized controlled trials. J. Med. Internet Res. **24**(6), e24111 (2022)
12. Demeco, A., Zola, L., Frizziero, A., Martini, C., Palumbo, A., Foresti, R., Buccino, G., Costantino, C.: Immersive virtual reality in post-stroke rehabilitation: a systematic review. Sensors (Basel). **23**(3), 1712 (2023)
13. Rutkowski, S., Rutkowska, A., Kiper, P., Jastrzebski, D., Racheniuk, H., Turolla, A., Szczegielniak, J., Casaburi, R.: Virtual reality rehabilitation in patients with chronic obstructive pulmonary disease: a randomized controlled trial. Int. J. Chron. Obstruct. Pulmon. Dis. **15**, 117–124 (2020)
14. Feitosa, J.A., Fernandes, C.A., Casseb, R.F., Castellano, G.: Effects of virtual reality-based motor rehabilitation: a systematic review of fMRI studies. J. Neural Eng. **19**(1) (2022)
15. Carl, E., Stein, A.T., Levihn-Coon, A., Pogue, J.R., Rothbaum, B., Emmelkamp, P., Asmundson, G.J.G., Carlbring, P., Powers, M.B.: Virtual reality exposure therapy for anxiety and related disorders: a meta-analysis of randomized controlled trials. J. Anxiety Disord. **61**, 27–36 (2019 Jan)
16. Maples-Keller, J.L., Bunnell, B.E., Kim, S., Rothbaum, B.O.: The use of virtual reality technology in the treatment of anxiety and other psychiatric disorders. Harv. Rev. Psychiatry. **25**(3), 103–113 (2017)
17. Donnelly, M.R., Reinberg, R., Ito, K.L., Saldana, D., Neureither, M., Schmiesing, A., Jahng, E., Liew, S.L.: Virtual reality for the treatment of anxiety disorders: a scoping review. Am. J. Occup. Ther. **75**(6), 7506205040 (2021)
18. Mevlevioğlu, D., Tabirca, S., Murphy, D.: Anxiety classification in virtual reality using biosensors: a mini scoping review. PLoS One. **18**(7), e0287984 (2023)
19. Wiebe, A., Kannen, K., Selaskowski, B., Mehren, A., Thöne, A.K., Pramme, L., Blumenthal, N., Li, M., Asché, L., Jonas, S., Bey, K., Schulze, M., Steffens, M., Pensel, M.C., Guth, M., Rohlfsen, F., Ekhlas, M., Lügering, H., Fileccia, H., Pakos, J., Lux, S., Philipsen, A., Braun, N.: Virtual reality in the diagnostic and therapy for mental disorders: a systematic review. Clin. Psychol. Rev. **98**, 102213 (2022)
20. Yen, H.Y., Chiu, H.L.: Virtual reality exergames for improving older adults' cognition and depression: a systematic review and meta-analysis of randomized control trials. J. Am. Med. Dir. Assoc. **22**(5), 995–1002 (2021)
21. Ioannou, A., Papastavrou, E., Avraamides, M.N., Charalambous, A.: Virtual reality and symptoms management of anxiety, depression, fatigue, and pain: a systematic review. SAGE Open Nurs. **6**, 2377960820936163 (2020)
22. Zhai, K., Dilawar, A., Yousef, M.S., Holroyd, S., El-Hammali, H., Abdelmonem, M.: Virtual reality therapy for depression and mood in long-term care facilities. Geriatrics (Basel). **6**(2), 58 (2021)
23. Huang, Q., Lin, J., Han, R., Peng, C., Huang, A.: Using virtual reality exposure therapy in pain management: a systematic review and meta-analysis of randomized controlled trials. Value Health. **25**(2), 288–301 (2022)
24. Tas, F.Q., van Eijk, C.A.M., Staals, L.M., Legerstee, J.S., Dierckx, B.: Virtual reality in pediatrics, effects on pain and anxiety: a systematic review and meta-analysis update. Paediatr. Anaesth. **32**(12), 1292–1304 (2022)
25. Grassini, S.: Virtual reality assisted non-pharmacological treatments in chronic pain management: a systematic review and quantitative meta-analysis. Int. J. Environ. Res. Public Health. **19**(7), 4071 (2022)

26. Jiang, H., Vimalesvaran, S., Wang, J.K., Lim, K.B., Mogali, S.R., Car, L.T.: Virtual reality in medical students' education: scoping review. JMIR Med. Educ. **8**(1), e34860 (2022). https://doi.org/10.2196/34860
27. Tudor Car, L., Kyaw, B.M., Teo, A., Fox, T.E., Vimalesvaran, S., Apfelbacher, C., Kemp, S., Chavannes, N.: Outcomes, measurement instruments, and their validity evidence in randomized controlled trials on virtual, augmented, and mixed reality in undergraduate medical education: systematic mapping review. JMIR Serious Games. **10**(2), e29594 (2022). https://doi.org/10.2196/29594
28. Tursø-Finnich, T., Jensen, R.O., Jensen, L.X., Konge, L., Thinggaard, E.: Virtual reality head-mounted displays in medical education: a systematic review. Simul. Healthc. **18**(1), 42–50 (2023)
29. Plotzky, C., Lindwedel, U., Sorber, M., Loessl, B., König, P., Kunze, C., Kugler, C., Meng, M.: Virtual reality simulations in nurse education: a systematic mapping review. Nurse Educ. Today. **101**, 104868 (2021)
30. Schuelke, S., Aurit, S., Connot, N., Denney, S.: Virtual nursing: the new reality in quality care. Nurs. Adm. Q. **43**(4), 322–328 (2019)
31. Zhao, J., Xu, X., Jiang, H., Ding, Y.: The effectiveness of virtual reality-based technology on anatomy teaching: a meta-analysis of randomized controlled studies. BMC Med. Educ. **20**(1), 127 (2020)
32. Moro, C., Birt, J., Stromberga, Z., Phelps, C., Clark, J., Glasziou, P., Scott, A.M.: Virtual and augmented reality enhancements to medical and science student physiology and anatomy test performance: a systematic review and meta-analysis. Anat. Sci. Educ. **14**(3), 368–376 (2021)
33. García-Robles, P., Cortés-Pérez, I., Nieto-Escámez, F.A., García-López, H., Obrero-Gaitán, E., Osuna-Pérez, M.C.: Immersive virtual reality and augmented reality in anatomy education: a systematic review and meta-analysis. Anat. Sci. Educ. (2024)
34. Ryan, G.V., Callaghan, S., Rafferty, A., Higgins, M.F., Mangina, E., McAuliffe, F.: Learning outcomes of immersive technologies in health care student education: systematic review of the literature. J. Med. Internet Res. **24**(2), e30082 (2022)
35. Martin, J.E., Tyndall, D.: Effect of manikin and virtual simulation on clinical judgment. J. Nurs. Educ. **61**(12), 693–699 (2022)
36. Mao, R.Q., Lan, L., Kay, J., Lohre, R., Ayeni, O.R., Goel, D.P., Sa, D.: Immersive virtual reality for surgical training: a systematic review. J. Surg. Res. **268**, 40–58 (2021)
37. Yi, W.S., Rouhi, A.D., Duffy, C.C., Ghanem, Y.K., Williams, N.N., Dumon, K.R.: A systematic review of immersive virtual reality for nontechnical skills training in surgery. J. Surg. Educ. **81**(1), 25–36 (2024)
38. Clarkson, L., Williams, M.: EMS Mass Casualty Triage StatPearis (2023). Available at https://www.ncbi.nlm.nih.gov/books/NBK459369/
39. Vincent, D.S., Sherstyuk, A., Burgess, L., Connolly, K.K.: Teaching mass casualty triage skills using immersive three-dimensional virtual reality. Acad. Emerg. Med. **15**(11), 1160–1165 (2008)
40. Andreatta, P.B., Maslowski, E., Petty, S., Shim, W., Marsh, M., Hall, T., Stern, S., Frankel, J.: Virtual reality triage training provides a viable solution for disaster-preparedness. Acad. Emerg. Med. **17**(8), 870–876 (2010)
41. Luigi Ingrassia, P., Ragazzoni, L., Carenzo, L., Colombo, D., Ripoll Gallardo, A., Della Corte, F.: Virtual reality and live simulation: a comparison between two simulation tools for assessing mass casualty triage skills. Eur J Emerg Med. **22**(2), 121–127 (2015)
42. Lowe, J., Peng, C., Winstead-Derlega, C., Curtis, H.: 360 virtual reality pediatric mass casualty incident: a cross sectional observational study of triage and out-of-hospital intervention accuracy at a national conference. J. Am. Coll. Emerg. Physicians Open. **1**(5), 974–980 (2020)
43. Hopper, L., Brook, L., Bartlett, D.: Virtual reality triage training can provide comparable simulation efficacy for paramedicine students compared to live simulation-based scenarios. Prehosp. Emerg. Care. **24**(4), 525–536 (2020)

44. Harada, S., Suga, R., Suzuki, K., Kitano, S., Fujimoto, K., Narikawa, K., Nakazawa, M., Ogawa, S.: Usefulness of self-selected scenarios for simple triage and rapid treatment method using virtual reality. J. Nippon Med. Sch. **91**(1), 99–107 (2024)
45. Farra, S.L., Smith, S.J., Ulrich, D.L.: The student experience with varying immersion levels of virtual reality simulation. Nurs. Educ. Perspect. **39**(2), 99–101 (2018)
46. Takei, Y., Nishi, T., Matsubara, H., Hashimoto, M., Inaba, H.: Factors associated with quality of bystander CPR: the presence of multiple rescuers and bystander-initiated CPR without instruction. Resuscitation. **85**, 492–498 (2014)
47. Chocron, R., Jobe, J., Guan, S., Kim, M., Shigemura, M., Fahrenbruch, C., Rea, T.: Bystander cardiopulmonary resuscitation quality: potential for improvements in cardiac arrest resuscitation. J. Am. Heart Assoc. **10**(6), e017930 (2021)
48. Talikowska, M., Tohira, H., Finn, J.: Cardiopulmonary resuscitation quality and patient survival outcome in cardiac arrest: a systematic review and meta-analysis. Resuscitation. **96**, 66–77 (2015)
49. Olasveengen, T.M., Semeraro, F., Ristagno, G., Castren, M., Handley, A., Kuzovlev, A., Monsieurs, K.G., Raffay, V., Smyth, M., Soar, J., Svavarsdottir, H., Perkins, G.D.: European Resuscitation Council Guidelines 2021: basic life support. Resuscitation. **161**, 98–114 (2021)
50. Riggs, M., Franklin, R., Saylany, L.: Associations between cardiopulmonary resuscitation (CPR) knowledge, self-efficacy, training history and willingness to perform CPR and CPR psychomotor skills: a systematic review. Resuscitation. **138**, 259–272 (2019)
51. Birkun, A.: Distant learning of BLS amid the COVID-19 pandemic: influence of the outbreak on lay trainees' willingness to attempt CPR, and the motivating effect of the training. Resuscitation. **152**, 105–106 (2020)
52. Cheng, A., Nadkarni, V.M., Mancini, M.B., Hunt, E.A., Sinz, E.H., Merchant, R.M., Donoghue, A., Duff, J.P., Eppich, W., Auerbach, M., Bigham, B.L., Blewer, A.L., Chan, P.S., Bhanji, F.: Resuscitation education science: educational strategies to improve outcomes from cardiac arrest: a scientific statement from the American Heart Association. Circulation. **138**(6), e82–e122 (2018)
53. Greif, R., Bhanji, F., Bigham, B.L., Bray, J., Breckwoldt, J., Cheng, A., Duff, J.P., Gilfoyle, E., Hsieh, M.J., Iwami, T., Lauridsen, K.G., Lockey, A.S., Ma, M.H., Monsieurs, K.G., Okamoto, D., Pellegrino, J.L., Yeung, J., Finn, J.C., Baldi, E., Beck, S., Beckers, S.K., Blewer, A.L., Boulton, A., Cheng-Heng, L., Yang, C.W., Coppola, A., Dainty, K.N., Damjanovic, D., Djärv, T., Donoghue, A., Georgiou, M., Gunson, I., Krob, J.L., Kuzovlev, A., Ko, Y.C., Leary, M., Lin, Y., Mancini, M.E., Matsuyama, T., Navarro, K., Nehme, Z., Orkin, A.M., Pellis, T., Pflanzl-Knizacek, L., Pisapia, L., Saviani, M., Sawyer, T., Scapigliati, A., Schnaubelt, S., Scholefield, B., Semeraro, F., Shammet, S., Smyth, M.A., Ward, A., Zace, D.: Education, implementation, and teams: 2020 international consensus on cardiopulmonary resuscitation and emergency cardiovascular care science with treatment recommendations. Resuscitation. **156**, A188–A239 (2020)
54. Greif, R., Lockey, A., Breckwoldt, J., Carmona, F., Conaghan, P., Kuzovlev, A., Pflanzl-Knizacek, L., Sari, F., Shammet, S., Scapigliati, A., Turner, N., Yeung, J., Monsieurs, K.G.: European Resuscitation Council Guidelines 2021: education for resuscitation. Resuscitation. **161**, 388–407 (2021)
55. Ricci, S., Calandrino, A., Borgonovo, G., Chirico, M., Casadio, M.: Viewpoint: virtual and augmented reality in basic and advanced life support training. JMIR Serious Games. **10**(1), e28595 (2022)
56. Jaskiewicz, F., Kowalewski, D., Starosta, K., Cierniak, M., Timler, D.: Chest compressions quality during sudden cardiac arrest scenario performed in virtual reality: a crossover study in a training environment. Medicine (Baltimore). **99**(48), e23374 (2020)
57. Issleib, M., Kromer, A., Pinnschmidt, H.O., Süss-Havemann, C., Kubitz, J.C.: Virtual reality as a teaching method for resuscitation training in undergraduate first year medical students: a randomized controlled trial. Scand. J. Trauma Resusc Emerg. Med. **29**(1), 27 (2021)

58. Moll-Khosrawi, P., Falb, A., Pinnschmidt, H., Zöllner, C., Issleib, M.: Virtual reality as a teaching method for resuscitation training in undergraduate first year medical students during COVID-19 pandemic: a randomised controlled trial. BMC Med. Educ. **22**(1), 483 (2022)
59. García Fierros, F.J., Moreno Escobar, J.J., Sepúlveda Cervantes, G., Morales Matamoros, O., Tejeida Padilla, R.: VirtualCPR: virtual reality mobile application for training in cardiopulmonary resuscitation techniques. Sensors (Basel). **21**(7), 2504 (2021)
60. Sadeghi, A.H., Peek, J.J., Max, S.A., Smit, L.L., Martina, B.G., Rosalia, R.A., Bakhuis, W., Bogers, A.J., Mahtab, E.A.: Virtual reality simulation training for cardiopulmonary resuscitation after cardiac surgery: face and content validity study. JMIR Serious Games. **10**(1), e30456 (2022)
61. Kim, E.A., Cho, K.J.: Comparing the effectiveness of two new CPR training methods in Korea: medical virtual reality simulation and flipped learning. Iran. J. Public Health. **52**(7), 1428–1438 (2023)
62. Lee, D.K., Choi, H., Jheon, S., Jo, Y.H., Im, C.W., Il, S.Y.: Development of an extended reality simulator for basic life support training. IEEE J. Transl. Eng. Health Med. **10**, 4900507 (2022)
63. Lau, Y., Nyoe, R.S.S., Wong, S.N., Ab Hamid, Z.B., Leong, B.S., Lau, S.T.: Effectiveness of digital resuscitation training in improving knowledge and skills: a systematic review and meta-analysis of randomised controlled trials. Resuscitation. **131**, 14–23 (2018)
64. Adams, A.J., Wasson, E.A., Admire, J.R., Pablo Gomez, P., Babayeuski, R.A., Sako, E.Y., Willis, R.E.: A comparison of teaching modalities and fidelity of simulation levels in teaching resuscitation scenarios. J. Surg. Educ. **72**(5), 778–785 (2015)
65. Boada, I., Rodriguez-Benitez, A., Garcia-Gonzalez, J.M., Olivet, J., Carreras, V., Sbert, M.: Using a serious game to complement CPR instruction in a nurse faculty. Comput. Methods Prog. Biomed. **122**(2), 282–291 (2015)
66. Bonnetain, E., Boucheix, J.M., Hamet, M., Freysz, M.: Benefits of computer screen-based simulation in learning cardiac arrest procedures. Med. Educ. **44**(7), 716–722 (2010)
67. Chung, C.H., Siu, A.Y., Po, L.L., Lam, C.Y., Wong, P.C.: Comparing the effectiveness of video self-instruction versus traditional classroom instruction targeted at cardiopulmonary resuscitation skills for laypersons: a prospective randomised controlled trial. Hong Kong Med. J. **16**(3), 165–170 (2010)
68. Delasobera, B.E., Goodwin, T.L., Strehlow, M., Gilbert, G., D'Souza, P., Alok, A., Raje, P., Mahadevan, S.V.: Evaluating the efficacy of simulators and multimedia for refreshing ACLS skills in India. Resuscitation. **81**(2), 217–223 (2010)
69. Einspruch, E., Lembach, J., Lynch, B., Lee, W., Harper, R., Fleischman, R.: Basic life support instructor training: comparison of instructor-led and self-guided training. J. Nurses Staff Dev. **27**, E4–E9 (2011)
70. Iserbyt, P., Charlier, N., Mols, L.: Learning basic life support (BLS) with tablet PCs in reciprocal learning at school: are videos superior to pictures? A randomized controlled trial. Resuscitation. **85**(6), 809–813 (2014)
71. Kononowicz, A.A., Krawczyk, P., Cebula, G., Dembkowska, M., Drab, E., Frączek, B., Stachoń, A.J., Andres, J.: Effects of introducing a voluntary virtual patient module to a basic life support with an automated external defibrillator course: a randomised trial. BMC Med. Educ. **12**, 41 (2012)
72. Krogh, L.Q., Bjørnshave, K., Vestergaard, L.D., Sharma, M.B., Rasmussen, S.E., Nielsen, H.V., Thim, T., Løfgren, B.: E-learning in pediatric basic life support: a randomized controlled non-inferiority study. Resuscitation. **90**, 7–12 (2015)
73. Lehmann, R., Thiessen, C., Frick, B., Bosse, H.M., Nikendei, C., Hoffmann, G.F., Tönshoff, B., Huwendiek, S.: Improving pediatric basic life support performance through blended learning with web-based virtual patients: randomized controlled trial. J. Med. Internet Res. **17**(7), e162 (2015)
74. Lippmann, J., Livingston, P., Craike, M.J.: Comparison of two modes of delivery of first aid training including basic life support. Health Educ. J. **2011**(70), 131–140 (2011)

75. Magura, S., Miller, M.G., Michael, T., Bensley, R., Burkhardt, J.T., Puente, A.C., Sullins, C.: Novel electronic refreshers for cardiopulmonary resuscitation: a randomized controlled trial. BMC Emerg. Med. **12**, 18 (2012)
76. Mancini, M.E., Cazzell, M., Kardong-Edgren, S., Cason, C.L.: Improving workplace safety training using a self-directed CPR-AED learning program. AAOHN J. **57**(4), 159–167; quiz 168–9 (2009)
77. Perkins, G.D., Fullerton, J.N., Davis-Gomez, N., Davies, R.P., Baldock, C., Stevens, H., Bullock, I., Lockey, A.S.: The effect of pre-course e-learning prior to advanced life support training: a randomised controlled trial. Resuscitation. **81**(7), 877–881 (2010)
78. Perkins, G.D., Kimani, P.K., Bullock, I., Clutton-Brock, T., Davies, R.P., Gale, M., Lam, J., Lockey, A., Stallard, N., Electronic Advanced Life Support Collaborators: Improving the efficiency of advanced life support training: a randomized, controlled trial. Ann. Intern. Med. **157**(1), 19–28 (2012)
79. Rehberg, R.S., Diaz, L.G., Middlemas, D.A.: Classroom versus computer-based CPR training: a comparison of the effectiveness of two instructional methods. Athl. Train. Educ. J. **4**(3), 98–103 (2009)
80. Roppolo, L.P., Heymann, R., Pepe, P., Wagner, J., Commons, B., Miller, R., Allen, E., Horne, L., Wainscott, M.P., Idris, A.H.: A randomized controlled trial comparing traditional training in cardiopulmonary resuscitation (CPR) to self-directed CPR learning in first year medical students: the two-person CPR study. Resuscitation. **82**(3), 319–325 (2011)
81. Mohd Saiboon, I., Jaafar, M.J., Ahmad, N.S., Nasarudin, N.M., Mohamad, N., Ahmad, M.R., Gilbert, J.H.: Emergency skills learning on video (ESLOV): a single-blinded randomized control trial of teaching common emergency skills using self-instruction video (SIV) versus traditional face-to-face (FTF) methods. Med. Teach. **36**(3), 245–250 (2014)
82. Saraç, L., Ok, A.: The effects of different instructional methods on students' acquisition and retention of cardiopulmonary resuscitation skills. Resuscitation. **81**(5), 555–561 (2010)
83. Weiner, G.M., Menghini, K., Zaichkin, J., Caid, A.E., Jacoby, C.J., Simon, W.M.: Self-directed versus traditional classroom training for neonatal resuscitation. Pediatrics. **127**(4), 713–719 (2011)
84. Alcázar Artero, P.M., Pardo Rios, M., Greif, R., Ocampo Cervantes, A.B., Gijón-Nogueron, G., Barcala-Furelos, R., Aranda-García, S., Ramos Petersen, L.: Efficiency of virtual reality for cardiopulmonary resuscitation training of adult laypersons: a systematic review. Medicine (Baltimore). **102**(4), e32736 (2023)
85. Semeraro, F., Ristagno, G., Giulini, G., Gnudi, T., Kayal, J.S., Monesi, A., Tucci, R., Scapigliati, A.: Virtual reality cardiopulmonary resuscitation (CPR): comparison with a standard CPR training mannequin. Resuscitation. **135**, 234–235 (2019)
86. McGrath, J.L., Taekman, J.M., Dev, P., Danforth, D.R., Mohan, D., Kman, N., Crichlow, A., Bond, W.F.: Using virtual reality simulation environments to assess competence for emergency medicine learners. Acad. Emerg. Med. **25**(2), 186–195 (2018)
87. Mahmood, F., Mahmood, E., Dorfman, R.G., Mitchell, J., Mahmood, F.U., Jones, S.B., Matyal, R.: Augmented reality and ultrasound education: initial experience. J. Cardiothorac. Vasc. Anesth. **32**(3), 1363–1367 (2018)
88. Huang, C.Y., Thomas, J.B., Alismail, A., Cohen, A., Almutairi, W., Daher, N.S., Terry, M.H., Tan, L.D.: The use of augmented reality glasses in central line simulation: "see one, simulate many, do one competently, and teach everyone". Adv. Med. Educ. Pract. **9**, 357–363 (2018)
89. Balian, S., McGovern, S.K., Abella, B.S., Blewer, A.L., Leary, M.: Feasibility of an augmented reality cardiopulmonary resuscitation training system for health care providers. Heliyon. **5**(8), e02205 (2019)
90. Berg, K.M., Bray, J.E., Ng, K.C., Liley, H.G., Greif, R., Carlson, J.N., Morley, P.T., Drennan, I.R., Smyth, M., Scholefield, B.R., Weiner, G.M., Cheng, A., Djärv, T., Abelairas-Gómez, C., Acworth, J., Andersen, L.W., Atkins, D.L., Berry, D.C., Bhanji, F., Bierens, J., Bittencourt Couto, T., Borra, V., Böttiger, B.W., Bradley, R.N., Breckwoldt, J., Cassan, P., Chang, W.T., Charlton, N.P., Chung, S.P., Considine, J., Costa-Nobre, D.T., Couper, K., Dainty, K.N.,

Dassanayake, V., Davis, P.G., Dawson, J.A., de Almeida, M.F., De Caen, A.R., Deakin, C.D., Dicker, B., Douma, M.J., Eastwood, K., El-Naggar, W., Fabres, J.G., Fawke, J., Fijacko, N., Finn, J.C., Flores, G.E., Foglia, E.E., Folke, F., Gilfoyle, E., Goolsby, C.A., Granfeldt, A., Guerguerian, A.M., Guinsburg, R., Hatanaka, T., Hirsch, K.G., Holmberg, M.J., Hosono, S., Hsieh, M.J., Hsu, C.H., Ikeyama, T., Isayama, T., Johnson, N.J., Kapadia, V.S., Kawakami, M.D., Kim, H.S., Kleinman, M.E., Kloeck, D.A., Kudenchuk, P., Kule, A., Kurosawa, H., Lagina, A.T., Lauridsen, K.G., Lavonas, E.J., Lee, H.C., Lin, Y., Lockey, A.S., Macneil, F., Maconochie, I.K., Madar, R.J., Malta Hansen, C., Masterson, S., Matsuyama, T., McKinlay, C.J.D., Meyran, D., Monnelly, V., Nadkarni, V., Nakwa, F.L., Nation, K.J., Nehme, Z., Nemeth, M., Neumar, R.W., Nicholson, T., Nikolaou, N., Nishiyama, C., Norii, T., Nuthall, G.A., Ohshimo, S., Olasveengen, T.M., Ong, Y.G., Orkin, A.M., Parr, M.J., Patocka, C., Perkins, G.D., Perlman, J.M., Rabi, Y., Raitt, J., Ramachandran, S., Ramaswamy, V.V., Raymond, T.T., Reis, A.G., Reynolds, J.C., Ristagno, G., Rodriguez-Nunez, A., Roehr, C.C., Rüdiger, M., Sakamoto, T., Sandroni, C., Sawyer, T.L., Schexnayder, S.M., Schmölzer, G.M., Schnaubelt, S., Semeraro, F., Singletary, E.M., Skrifvars, M.B., Smith, C.M., Soar, J., Stassen, W., Sugiura, T., Tijssen, J.A., Topjian, A.A., Trevisanuto, D., Vaillancourt, C., Wyckoff, M.H., Wyllie, J.P., Yang, C.W., Yeung, J., Zelop, C.M., Zideman, D.A., Nolan, J.P., Collaborators: 2023 international consensus on cardiopulmonary resuscitation and emergency cardiovascular care science with treatment recommendations: summary from the basic life support; advanced life support; pediatric life support; neonatal life support; education, implementation, and teams; and first aid task forces. Circulation. **148**(24), e187–e280 (2023)

91. Flin, R., Maran, N.: Identifying and training non-technical skills for teams in acute medicine. Qual. Saf. Health Care. **13**(Suppl 1), 80–84 (2004)
92. Frenk, J., Chen, L., Bhutta, Z.A., Cohen, J., Crisp, N., Evans, T., Fineberg, H., Garcia, P., Ke, Y., Kelley, P., Kistnasamy, B., Meleis, A., Naylor, D., Pablos-Mendez, A., Reddy, S., Scrimshaw, S., Sepulveda, J., Serwadda, D., Zurayk, H.: Health professionals for a new century: transforming education to strengthen health systems in an interdependent world. Lancet. **376**(9756), 1923–1958 (2010)
93. Cheng, A., Nadkarni, V.M., Mancini, M.B., Hunt, E.A., Sinz, E.H., Merchant, R.M., Donoghue, A., Duff, J.P., Eppich, W., Auerbach, M., Bigham, B.L., Blewer, A.L., Chan, P.S., Bhanji, F.: Resuscitation education science: educational strategies to improve outcomes from cardiac arrest: a scientific statement from the American Heart Association. Circulation. **138**(6), 82–122 (2018)
94. Moran, C.G., Lecky, F., Bouamra, O., Lawrence, T., Edwards, A., Woodford, M., Willett, K., Coats, T.J.: Changing the system—major trauma patients and their outcomes in the NHS (England) 2008–17. EClinicalMedicine. **2–3**, 13–21 (2018)
95. ATLS Subcommittee; American College of Surgeons' Committee on Trauma; International ATLS working group: Advanced trauma life support (ATLS®): the ninth edition. J. Trauma Acute Care Surg. **74**(5), 1363–1366 (2013)
96. Advanced Trauma Life Support: https://www.facs.org/quality-programs/trauma/education/advanced-trauma-life-support/
97. Advanced Trauma Life Support® (ATLS®): https://www.rcseng.ac.uk/education-and-exams/courses/search/advanced-trauma-life-support-atls-provider-programme/
98. JATEC: https://www.jtcr-jatec.org/index_jatec.html
99. Hainsworth, L., Kosti, A., Lloyd, A., Kiddle, A., Bamford, R., Hunter, I.: Teaching the management of trauma patients through virtual reality. Ann. R. Coll. Surg. Engl. **104**(5), 330–333 (2022)
100. Holla, M., Berg, M.V.D.: Virtual reality techniques for trauma education. Injury. **53**(Suppl 3), S64–S68 (2022)
101. Liaw, S.Y., Tan, J.Z., Lim, S., Zhou, W., Yap, J., Ratan, R., Ooi, S.L., Wong, S.J., Seah, B., Chua, W.L.: Artificial intelligence in virtual reality simulation for interprofessional communication training: mixed method study. Nurse Educ. Today. **122**, 105718 (2023)

102. Liaw, S.Y., Zhou, W.T., Lau, T.C., Siau, C., Chan, S.W.: An interprofessional communication training using simulation to enhance safe care for a deteriorating patient. Nurse Educ. Today. **34**(2), 259–264 (2014)
103. Chang, C.W., Lin, C.W., Huang, C.Y., Hsu, C.W., Sung, H.Y., Cheng, S.F.: Effectiveness of the virtual reality chemical disaster training program in emergency nurses: a quasi experimental study. Nurse Educ. Today. **119**, 105613 (2022)
104. Regal, G., Pretolesi, D., Schrom-Feiertag, H., Puthenkalam, J., Migliorini, M., De Maio, E., Scarrone, F., Nadalin, M., Guarneri, M., Xerri, G.P., et al.: Challenges in virtual reality training for CBRN events. Multimodal Technol. Interact. **7**(9), 88 (2023)
105. Kako, M., Hammad, K., Mitani, S., Arbon, P.: Existing approaches to chemical, biological, radiological, and nuclear (CBRN) education and training for health professionals: findings from an integrative literature review. Prehosp. Disaster Med. **33**, 1–9 (2018)
106. De Lorenzis, F., Pratticò, F.G., Cultrera, M., Migliorini, M., Lamberti, F.: An immersive virtual reality training environment for CBRN procedures. In: Methodologies and Use Cases on Extended Reality for Training and Education, pp. 73–96. IGI Global, Hershey (2022)
107. Savir, S., Khan, A.A., Yunus, R.A., Rehman, T.A., Saeed, S., Sohail, M., Sharkey, A., Mitchell, J., Matyal, R.: Virtual reality: the future of invasive procedure training? J. Cardiothorac. Vasc. Anesth. **37**(10), 2090–2097 (2023)
108. Liaw, S.Y., Chan, S.W., Chen, F., Hooi, S.C., Siau, C.: Comparison of virtual patient simulation with mannequin-based simulation for improving clinical performances in assessing and managing clinical deterioration: randomized controlled trial. J. Med. Internet Res. **16**(9), e214 (2014)
109. Barsom, E.Z., Graafland, M., Schijven, M.P.: Systematic review on the effectiveness of augmented reality applications in medical training. Surg. Endosc. **30**, 4174–4183 (2016)
110. Ahlberg, G., Enochsson, L., Gallagher, A.G., Hedman, L., Hogman, C., McClusky 3rd, D.A., Ramel, S., Smith, C.D., Arvidsson, D.: Proficiency-based virtual reality training significantly reduces the error rate for residents during their first 10 laparoscopic cholecystectomies. Am. J. Surg. **193**(6), 797–804 (2007)
111. Schirmer, C.M., Elder, J.B., Roitberg, B., Lobel, D.A.: Virtual reality-based simulation training for ventriculostomy: an evidence-based approach. Neurosurgery. **73**(Suppl 1), 66–73 (2013)
112. İsmailoğlu, E.G., Zaybak, A.: Comparison of the effectiveness of a virtual simulator with a plastic arm model in teaching intravenous catheter insertion skills. Comput Inform. Nurs. **36**(2), 98–105 (2018)
113. Smith, S.J., Farra, S.L., Ulrich, D.L., Hodgson, E., Nicely, S., Mickle, A.: Effectiveness of two varying levels of virtual reality simulation. Nurs. Educ. Perspect. **39**(6), E10–E15 (2018)
114. Van der Meijden, O.A., Schijven, M.P.: The value of haptic feedback in conventional and robot-assisted minimal invasive surgery and virtual reality training: a current review. Surg. Endosc. **23**(6), 1180–1190 (2009)
115. Shi, Y., Shen, G.: Haptic sensing and feedback techniques toward virtual reality. Research (Wash D C). **7**, 0333 (2024)

Chapter 8
General Information About Virtual Reality Headsets

Dana M. Barry and Hideyuki Kanematsu

Abstract This chapter presents general information about virtual reality (VR) headsets. It describes what they are and how they work. A virtual reality headset is a head-mounted display that creates a reality-based virtual environment. It includes software that allows individuals to interact with real or imaginary computer-simulated environments. These three-dimensional (3-D) experiences include both visual and auditory components. The headsets are used for VR video games, simulations, teaching purposes, and more. This chapter mentions a few benefits and limitations of VR headsets. It also presents and describes several studies in education showing that students who used the VR headsets had improvements in posttests, positive responses for the usefulness of the VR headsets, and experiences that were exciting, enjoyable, and motivating.

Keywords Components of virtual reality (VR) headsets · 3-D experiences · Uses of VR headsets · Benefits and limitations of VR headsets

8.1 Introduction

This chapter begins by introducing general information about virtual reality (VR) headsets, explaining their features and functionality. It also explores the diverse applications of VR headsets. Furthermore, the chapter highlights and evaluates several educational studies, which demonstrate that students using VR headsets showed

D. M. Barry (✉)
Clarkson University, Potsdam, NY, USA
e-mail: dbarry@clarkson.edu

H. Kanematsu
National Institute of Technology, Suzuka College, Suzuka, Mie, Japan
e-mail: hideyuki.kanematsu@bioenglab.org

D. M. Barry, H. Kanematsu (eds.), *Applications of Metaverse and Virtual Reality to Creative Education and Industry*, Intelligent Systems Reference Library 267, https://doi.org/10.1007/978-981-96-3341-8_8

Fig. 8.1 Below is a photo of an Oculus VR headset, which is placed over the user's head to immerse him or herself into a 3D virtual experience

improved posttest results, found the technology useful, and had engaging, enjoyable, and motivational experiences.

A VR headset is essentially a head-mounted device that generates a virtual environment within a reality. Figure 8.1 [1] shows an example of an Oculus VR headset. These headsets are equipped with software enabling users to engage in realistic and imaginative computer-simulated environments. The three-dimensional (3-D) simulations offered by these headsets include both visual and auditory elements. A primary objective of VR is to deliver an immersive, interactive setting that fully engages the user's senses, creating the perception of being physically present in a different world [2–20]. To enhance immersion, VR headsets are designed with a broad field of view to minimize outside distractions. These devices utilize advanced technology, integrating various components to craft a highly immersive and interactive experience. Key elements of this technology are briefly outlined to enhance the reader's understanding of how VR headsets operate [21–40].

Key Components Commonly Found in VR Headsets

Display: The display is responsible for delivering the visual output that forms the virtual environment. VR headsets typically employ high-resolution Liquid Crystal Display (LCD) screens to render lifelike graphics. These displays are divided into two segments, one for each eye, to achieve a 3D visual effect. The refresh rate is a crucial aspect, determining how frequently the display updates each second. Another vital parameter is the field of view (FOV), which dictates the extent of the visible virtual environment. To address motion sickness, display technology incorporates techniques such as low persistence, which reduces motion blur caused by fast-moving visuals, and positional tracking, which ensures that the virtual environment accurately aligns with a user's head movements.

Lenses: Positioned in front of the display, lenses focus and reshape the images to adjust for the distance between the eyes and achieve a broad field of view. Some VR headsets feature specialized lens technologies, such as Fresnel lenses, to enhance visual clarity and immersion.

Head Tracking Sensors: These sensors, including gyroscopes, accelerometers, and infrared devices, monitor the rotational and positional movements of the headset. This ensures that the virtual environment reacts realistically to the user's head movements.

Audio: Audio plays a crucial role in creating an immersive experience by enhancing the sense of presence and realism. Most VR headsets include built-in headphones or speakers that deliver sound directly to the user. Using 3D sound positioning, VR systems simulate directional and spatial audio effects, making sounds seem to originate from specific locations. Real-time audio processing adjusts sounds dynamically based on the user's head orientation, ensuring synchronization with their perspective.

Controllers and Input Devices: VR systems feature a range of controllers and input devices, such as handheld controllers and motion sensors, to enable users to interact with the virtual world. These devices allow users to manipulate objects, navigate menus, and perform actions. Motion controllers include built-in sensors to detect the position and movement of hands and fingers. Eye tracking is a relatively new technology that monitors a user's gaze, facilitating more natural interactions and inferring focus and attention. Voice control is also being incorporated into VR systems, enabling users to issue commands and interact using spoken input.

Connectivity: VR headsets typically connect to a computer or gaming device to access processing power, content libraries, and storage. Most headsets rely on cables, such as USB cords, for data transmission and connectivity with external devices.

Software and Content: The software and content are integral to delivering immersive experiences, such as games and videos. The software integrates with the headset's sensors and processing units to generate the virtual environment. By combining advanced technologies and precise calibrations, VR headsets create highly immersive experiences that replicate reality.

8.2 Uses for Virtual Reality (VR) Headsets

VR headsets have existing and potential applications in various fields including gaming, other forms of entertainment, virtual travel and tourism, product design, education, and more [21, 41–76]. Some of these uses are presented and briefly described.

Entertainment: Gaming

VR gaming has become increasingly popular in recent years because VR headsets immerse players into the game world by eliminating external distractions. This growth is also driven by the accessibility of VR headsets and significant investments by game developers in VR technology. As a result, many users have purchased VR devices to play trending games. A study investigated the impact of VR headsets on the gaming experience of video game players. Participants played the same game, *Half-Life 2*, using both a traditional computer monitor and the Oculus Rift, a VR headset [77]. Game user satisfaction was assessed using the psychometrically validated Game User Experience Satisfaction Scale (GUESS), which measures nine constructs of video game satisfaction. The findings showed that VR headsets enhanced enjoyment, creativity, and overall satisfaction. The study involved 40 undergraduate students (16 males and 24 females) aged between 18 and 40. Beyond gaming, VR headsets are also used to watch 360-degree movies, shows, and videos, enabling users to move and explore within immersive virtual worlds.

Entertainment: Live Music

Live music is emerging as a major application for VR, allowing users to attend concerts from anywhere in the world. Pre-recorded concerts are already available as VR experiences, with videos filmed in 360 degrees to make users feel as though they are physically at the event. This is particularly beneficial for individuals who cannot travel or afford concert tickets.

Entertainment: Virtual Travel and Tourism

Virtual travel and tourism provide immersive experiences by simulating real-world landscapes and locations. These experiences are ideal for individuals lacking the time or resources to visit distant places physically. Examples include virtual tours of museums or renowned destinations such as the Grand Canyon and Japanese temples.

VR Simulations: Medical Training

Immersive VR simulations use head-mounted displays (HMDs) to fully engage users in realistic training environments. These simulations allow medical professionals to practice procedures before performing them on actual patients, leading to improved patient outcomes. Medical students can use these virtual environments to learn various procedures interactively. Scenarios can be programmed to prepare students for different medical situations, offering step-by-step guidance and hands-on experience in performing specific surgeries.

VR Simulations: Driving

VR simulations are valuable for driver education and automotive research. They provide training opportunities for young drivers, allowing them to recognize and correct bad habits. Risky driving situations can be simulated without real-world dangers, and feedback is provided afterward to improve behavior. Additionally, VR driving simulations collect real-time data on user reactions, which can be used to design safer vehicles, develop training programs, and advance autonomous vehicle (AV) technology.

VR Simulations: Product Design

VR enables designers to visualize 3D models in virtual spaces, offering a cost-effective way to test and evaluate preliminary designs and prototypes. This eliminates the need for significant resources spent on physical prototypes. Virtual environments also allow designers to simulate how products react under various conditions, identifying areas for improvement before moving to real-world production.

VR Socializing: Social Interaction
VR provides platforms for individuals to communicate over long distances by creating realistic environments for interaction. Users can have avatars and engage with others as if meeting face-to-face. This was especially popular during the COVID-19 pandemic when travel restrictions and social distancing limited in-person interactions. VR headsets facilitated realistic social experiences, allowing users to attend events and participate in various engaging activities virtually.

8.3 Educational Research on VR Headset Applications

This section presents and briefly describes several studies on the use of VR headsets in educational settings. Although VR is not a new concept, advancements in immersive technologies have significantly increased its appeal among researchers. Modern VR head-mounted displays (HMDs), such as Oculus Rift, provide users with a highly immersive experience, creating the illusion of being in a virtual environment. Additionally, affordable HMDs like Google Cardboard enable users to explore and interact with immersive virtual environments. Figure 8.2 illustrates a sixth-grade student using a VR cardboard headset to study ocean life [78].

One study examining VR headsets in education involved 99 first-year psychology students at the University of Warwick in the UK [79]. The students were randomly assigned to one of three learning methods: traditional textbook-based learning, VR headset-based learning, and a two-dimensional video approach. All three methods used the same textual content and a three-dimensional model of a

Fig. 8.2 Below is a sixth-grade student using a VR cardboard headset to learn about ocean life

plant cell. The VR group utilized HTC VIVE headsets, which allowed participants to view and interact with the plant cell model. Students could highlight individual cell parts, resize the model, and explore a virtual room where the plant cell appeared as a floating object. A menu offered additional interactive options. The video approach provided a two-dimensional recording with text and graphics resembling the VR model, while the textbook approach used screenshots of the 3D model alongside the text. Pre- and post-knowledge tests, as well as questionnaires, were used to collect data. Students using VR headsets and the textbook method demonstrated better knowledge acquisition and understanding compared to those in the video group. The VR group performed best in memory retention and exhibited increased positive emotions, decreased negative emotions, and higher engagement. Conversely, the other groups showed a decline in positive emotions.

A pilot study was conducted with social work students to explore the impact of a 360-degree VR simulation [80]. This immersive tool, developed by the authors, introduced students to a typical New York City neighborhood. The simulation used recorded video captured with a multidirectional camera to create a 360-degree view of the urban environment. The study aimed to help students understand how location and resources affect community members. Thirty first-year graduate social work students participated, using Google Daydream headsets connected to mobile devices. Data were gathered through pretests, posttests, and surveys measuring students' perceptions and attitudes. Results showed significant posttest improvements and positive feedback on the simulation's usefulness, with participants describing the experience as enjoyable, engaging, and motivating.

Another project evaluated the effectiveness of two types of VR systems for teaching high school mathematics [81]. The study focused on solving linear equations with three variables and involved two student groups using different VR systems. The first group utilized a desktop-based VR system with HTC VIVE headsets and handheld controllers. The second group tested an all-in-one VR system using HTC VIVE Focus devices, which required students to remain seated. Participants completed pre- and post-questionnaires, and post-experiment interviews were conducted with both students and teachers for qualitative and quantitative data. Findings indicated that VR technology enhances the teaching of high school mathematics when combined with traditional methods. The desktop-based VR system proved more effective than the all-in-one system, as it offered higher display efficiency and better user control.

Furthermore, VR has been applied to support and enhance fieldwork in Geography [82]. Virtual field trips (VFTs) and fieldwork courses were developed using internet-based applications and software, with students primarily interacting via a computer monitor. In contrast, standalone VR headsets provide opportunities to further improve virtual fieldwork by offering an uninterrupted field of view and creating a sense of immersion and presence at specific field sites. VR serves as an excellent supplement for fieldwork preparation, equipping students with essential geographical skills before they engage in actual field studies. This method was used to prepare first-year geography students for a residential field trip to Snowdonia National Park in Wales. Figure 8.3 shows a photo of this park's landscape [83].

Fig. 8.3 This photo is a display of Snowdonia National Park

In addition to exploring the park's natural scenery, the project included a focus on social and economic issues within the park. Before commencing the fieldwork, the students attended preparatory lectures and used Oculus Go headsets to observe and explore the landscape virtually. (Footage of Snowdonia National Park was downloaded onto the headsets to help students familiarize themselves with the site.) Using the VR headsets, which provided a 360-degree perspective, students worked in pairs to analyze the landscape by identifying wildlife, assessing risks and hazards, examining roads and vehicles, and noting uneven terrain. They found this collaborative research project enjoyable and used the insights gained during their discussions to enhance their understanding of the field site prior to their visit. This engaging activity also encouraged the development of critical and analytical thinking skills. Figure 8.3 can be reviewed for general information about the park's landscape.

8.4 Advantages and Challenges of VR Headsets

Virtual reality technology offers significant potential for diverse applications, including engineering, education, medicine, entertainment, and beyond. This potential stems from the advanced integration of programmed software and hardware. An additional advantage is the availability of affordable head-mounted displays, such as Google Cardboard, which allow users from various backgrounds to experience immersive virtual environments.

VR headsets generate virtual environments that enable users to interact with them. One of their primary objectives is to create an immersive and interactive experience that involves the senses, making users feel as though they are present in an alternate reality. These headsets are designed with a wide field of view to minimize external distractions. They are also valuable tools for virtual tours, providing an excellent option for individuals who lack the time or resources to travel to distant locations. It is also useful for simulations that provide driver training, industrial opportunities for product design, and an opportunity for medical professionals to practice procedures before operating on a patient, etc.

In addition to benefits, VR headsets have some limitations. As the technology develops, more progress must occur before VR can be fully used in all possible applications. Another issue is that VR headsets tend to be heavy and cause individuals to have headaches and pain, especially in the neck and shoulders. An important problem, because of using a VR headset, is cybersickness. This is where users experience nausea and dizziness like motion sickness. In addition, a person may become addicted to using the VR headset and isolate him or herself from society.

8.5 Conclusions

This chapter provides general information about virtual reality (VR) headsets. It starts by describing what they are and how they work. A VR headset is a head-mounted display that creates a reality-based virtual environment. It includes software that allows individuals to interact with real or imaginary computer-simulated environments. These three-dimensional experiences include both visual and auditory components. Some uses for VR headsets are presented. They include social interaction, various forms of entertainment like games, live music, virtual travel, and simulations for medical training, driver training, and product design. Several benefits and limitations for VR headsets are also mentioned. In addition, the chapter presents a few studies in education that show students who used VR headsets had improvements in posttests, positive responses for the usefulness of the VR headsets, and experiences that were exciting, enjoyable, and motivating.

References

1. Syced: File: Oculus Quest. jpeg License: Creative Commons CCO 1.O, Universal Public Domain (2019). File: Oculus Quest.jpeg—Wikimedia Commons
2. Al Awadhi, et al.: Virtual reality application for interactive and informative learning. In: 2nd International Conference on Bioengineering for Smart Technologies. IEEE, pp. 1–4 (2017)
3. Biocca, F., Delaney, B.: Immersive Virtual Reality Technology Communication in the Age of Virtual Reality, pp. 57–124. Lawrence Erlbaum Associates, Inc, Hillsdale (1995)
4. Bowman, D.A., McMahan, R.P.: Virtual reality: how much immersion is enough? Computer. **40**(7), 36–43 (2007)

5. Carruth, D.W.: Virtual reality for education and workforce training 15th international conference on emerging eLearning technologies and applications, pp. 1–6. IEEE (2017)
6. Chavez, B., Bayona, S.: Virtual Reality in the Learning Process: Trends and Advances in Information Systems and Technologies, pp. 1345–1356. Springer (2018)
7. Chin, N., Gupte, A., Nguyen, J., Sukhin, S., Wang, G., Mirizio, J.: Using virtual reality for an immersive experience in the water cycle. In: IEEE MIT Undergraduate Research Technology Conference, pp. 1–4. IEEE (2017)
8. Dolezal, M., Chmelik, J., Liarokapis, F.: An Immersive Virtual Environment for Collaborative Geo-Visualization, pp. 272–275. VS-GAMES (2017)
9. Farra, S.L., Smith, S.J., Ulrich, D.L.: The student experience with varying immersion levels of virtual reality simulation. Nurs. Educ. Perspect. **39**(2), 99–101 (2018)
10. Feng, Z., González, V.A., Amor, R., Lovreglio, R., Cabrera-Guerrero, G.: Immersive virtual reality serious games for evacuation training and research: a systematic literature review. Comput. Educ. **127**, 252–266 (2018)
11. Gerloni, I.G., Carchiolo, V., Vitello, F.R., Sciacca, E., Becciani, U., Costa, A., et al.: Immersive Virtual Reality for Earth Sciences Federated Conference on Computer Science and Information System. IEEE, pp. 527–534 (2018)
12. Harrington, C.M., Kavanagh, D.O., Quinlan, J.F., Ryan, D., Dicker, P., O'Keeffe, D., et al.: Development and evaluation of a trauma decision-making simulator in oculus virtual reality. Am. J. Surg. **215**(1), 42–47 (2018)
13. Jensen, L., Konradsen, F.: A review of the use of virtual reality head-mounted displays in education and training. Educ. Inf. Technol. **23**(4), 1515–1529 (2018). https://doi.org/10.1007/s10639-017-9676-0
14. Kwon, B., Kim, J., Lee, K., Lee, Y.K., Park, S., Lee, S.: Implementation of a virtual training simulator based on 360° multi-view human action recognition. IEEE Access. **5**, 12496–12511 (2017). https://doi.org/10.1109/ACCESS.2017.2723039
15. Martín-Gutiérrez, J., Mora, C.E., Añorbe Díaz, B., González-Marrero, A.: Virtual technologies trends in education. EURASIA J. Math. Sci. Technol. Educ. **13**(2), 469–486 (2017)
16. Merchant, Z., Goetz, E.T., Cifuentes, L., Keeney-Kennicutt, W., Davis, T.J.: Effectiveness of virtual reality-based instruction on students' learning outcomes in k-12 and higher education: a meta-analysis. Comput. Educ. **70**, 29–40 (2014)
17. Papachristos, N.M., Vrellis, I., Mikropoulos, T.A.: A comparison between Oculus rift and a low-cost smartphone VR headset: immersive user experience and learning. In: IEEE 17th International Conference on Advanced Learning Technologies, pp. 477–481 (2017). https://doi.org/10.1109/ICALT.2017.145
18. Parong, J., Mayer, R.E.: Learning science in immersive virtual reality. J. Educ. Psychol. **110**(6), 785–797 (2018)
19. Smith, S.J., Farra, S.L., Ulrich, D.L., Hodgson, E., Nicely, S., Mickle, A.: Effectiveness of two varying levels of virtual reality simulation. Nurs. Educ. Perspect. **39**(6), E10–E15 (2018)
20. Suh, A., Prophet, J.: The state of immersive technology research: a literature analysis. Comput. Hum. Behav. **86**, 77–90 (2018)
21. Wang, P., Wu, P., Wang, J., Chi, H.-L., Wang, X.: A critical review of the use of virtual reality in construction engineering education and training. Int. J. Environ. Res. Public Health. **15**(6), 1204 (2018). https://doi.org/10.3390/ijerph15061204
22. Ye, Q., Hu, W., Zhou, H., Lei, Z., Guan, S.: VR interactive feature of html5-based webvr control laboratory by using head-mounted displays. Int. J. Online Eng. **14**(3) (2018)
23. Wohlgenannt, I., Simons, A., Stieglitz, S.: Virtual Reality. Bus. Inf. Syst. Eng. **62**, 455–461 (2020). https://doi.org/10.1007/s12599-020-00658-9
24. Rahouti, A., Lovreglio, R., Datoussaïd, S., Descamps, T.: Prototyping and validating a non-immersive virtual reality serious game for Healthcare Fire Safety Training. Fire. Technol. **57**, 3041–3078 (2021)
25. Slater, M., Sanchez-Vives, M.V.: Transcending the self in immersive virtual reality. Computer. **47**, 24–30 (2014)

26. Harley, D.: Palmer Luckey and the rise of contemporary virtual reality. Convergence. **26**, 1144–1158 (2019)
27. Velev, D., Zlateva, P.: Virtual reality challenges in education and training. Int. J. Learn. Teach. **3**, 33–37 (2017)
28. Riegler, A., Riener, A., Holzmann, C.: Virtual reality driving simulator for user studies on automated driving. In: Proceedings of the 11th International Conference on Automotive User Interfaces and Interactive Vehicular Applications: Adjunct Proceedings; Seattle, WA, USA. 17–19 September; pp. 502–507 (2019)
29. Walch, M., Frommel, J., Rogers, K., Schüssel, F., Hock, P., Dobbelstein, D., Weber, M.: Evaluating VR driving simulation from a player experience perspective. In: Proceedings of the 2017 CHI Conference Extended Abstracts on Human Factors in Computing Systems. Denver, CO, USA. 6–11 May 2017, pp. 2982–2989 (2017)
30. Lele, A.: Virtual reality and its military utility. J. Ambient. Intell. Humaniz. Comput. **4**, 17–26 (2011)
31. Ihemedu-Steinke, Q.C., Erbach, R., Halady, P., Meixner, G., Weber, M.: Virtual reality driving simulator based on head-mounted displays. In: Automotive User Interfaces, pp. 401–428. Springer, Cham (2017)
32. Morra, L., Lamberti, F., Prattico, F.G., La Rosa, S., Montuschi, P.: Building trust in autonomous vehicles: role of virtual reality driving simulators in HMI design. IEEE Trans. Veh. Technol. **68**, 9438–9450 (2019)
33. Christopoulos, A., Conrad, M., Shukla, M.: Increasing student engagement through virtual interactions: how? Virtual Reality. **22**, 353–369 (2018)
34. Radianti, J., Majchrzak, T.A., Fromm, J., Wohlgenannt, I.: A systematic review of immersive virtual reality applications for higher education: design elements, lessons learned, and research agenda. Comput. Educ. **147**, 103778 (2020). https://doi.org/10.1016/j.compedu.2019.103778
35. Lau, K.W., Lee, P.Y.: The use of virtual reality for creating unusual environmental stimulation to motivate students to explore creative ideas. Interact. Learn. Environ. **23**, 3–18 (2012)
36. Alhalabi, W.: Virtual reality systems enhance students' achievements in engineering education. Behav. Inform. Technol. **35**, 919–925 (2016)
37. Gadelha, R.: Revolutionizing Education: the promise of virtual reality. Child. Educ. **94**, 40–43 (2018)
38. Au, E.H., Lee, J.J.: Virtual reality in education: a tool for learning in the experience age. Int. J. Innov. Educ. **4**, 215 (2017). https://doi.org/10.1504/IJIIE.2017.091481
39. O'Connor, E.A., Domingo, J.: A practical guide, with theoretical underpinnings, for creating effective virtual reality learning environments. J. Educ. Technol. Syst. **45**, 343–364 (2017)
40. Hamilton, D., McKechnie, J., Edgerton, E., Wilson, C.: Immersive virtual reality as a pedagogical tool in education: a systematic literature review of quantitative learning outcomes and experimental design. J. Comput. Educ. **8**, 1–32 (2020)
41. Hoffmann, M., Meisen, T., Jeschke, S.: Shifting virtual reality education to the next level: experiencing remote laboratories through mixed reality. In: Engineering Education 4.0, pp. 235–249. Springer, Cham (2016)
42. Chang, X.-Q., Zhang, D.-H., Jin, X.-X.: Application of virtual reality technology in distance learning. Int. J. Emerg. Technol. Learn. **11**, 76–79 (2016)
43. Morozov, M., Gerasimov, A., Fominykh, M., Smorkalov, A.: Asynchronous immersive classes in a 3D virtual world: extended description of vAcademia. In: Transactions on Computational Science XVIII, vol. 7848, pp. 81–100. Springer, Cham (2013)
44. Nickel, F., Brzoska, J.A., Gondan, M., Rangnick, H.M., Chu, J., Kenngott, H.G., Linke, G.R., Kadmon, M., Fischer, L., Müller-Stich, B.P.: Virtual reality training versus blended learning of laparoscopic cholecystectomy. Medicine. **94**, e764 (2015)
45. Pottle, J.: Virtual reality and the transformation of medical education. Future Health J. **6**, 181–185 (2019)

46. Izard, S.G., Juanes, J.A., García-Peñalvo, F.J., Gonçalvez Estella, J.M., Ledesma, M.J.S., Ruisoto, P.: Virtual reality as an educational and training tool for medicine. J. Med. Syst. **42**, 50 (2018)
47. Willaert, W.I.M., Aggarwal, R., Van Herzeele, I., Cheshire, N.J., Vermassen, F.E.: Recent advancements in medical simulation: patient-Specific virtual reality simulation. World J. Surg. **36**, 1703–1712 (2012)
48. Javaid, M., Haleem, A.: Virtual reality applications toward medical field. Clin. Epidemiol. Glob. Health. **8**, 600–605 (2020)
49. Rizzo, A.S., Lange, B., Suma, E.A., Bolas, M.: Virtual reality and interactive digital game technology: new tools to address obesity and diabetes. J. Diabetes Sci. Technol. **5**, 256–264 (2011)
50. Segura-Ortí, E., García-Testal, A.: Intradialytic virtual reality exercise: increasing physical activity through technology. Semin. Dial. **32**, 331–335 (2019)
51. Pasco, D.: The potential of using virtual reality technology in physical activity settings. Quest. **65**, 429–441 (2013)
52. Bird, J.M.: The use of virtual reality head-mounted displays within applied sport psychology. J. Sport Psychol. Action. **11**, 115–128 (2020)
53. Srivastava, K., Chaudhury, S., Das, R.: Virtual reality applications in mental health: challenges and perspectives. Ind. Psychiatry J. **23**, 83–85 (2014)
54. Seabrook, E., Kelly, R., Foley, F., Theiler, S., Thomas, N., Wadley, G., Nedeljkovic, M.: Understanding how virtual reality can support mindfulness practice: mixed methods study. J. Med. Internet Res. **22**, e16106 (2020)
55. Kaplan-Rakowski, R., Johnson, K.R., Wojdynski, T.: The impact of virtual reality meditation on college students' exam performance. Smart Learn. Environ. **8**, 21 (2021)
56. Tarrant, J., Jackson, R., Viczko, J.: A feasibility test of a brief mobile virtual reality meditation for frontline healthcare workers in a hospital setting. Front. Virtual Real. **3**, 764745 (2022). https://doi.org/10.3389/frvir.2022.764745
57. Didehbani, N., Allen, T., Kandalaft, M., Krawczyk, D., Chapman, S.: Virtual reality social cognition training for children with high functioning autism. Comput. Hum. Behav. **62**, 703–711 (2016)
58. Kandalaft, M.R., Didehbani, N., Krawczyk, D.C., Allen, T.T., Chapman, S.B.: Virtual reality social cognition training for young adults with high-functioning Autism. J. Autism Dev. Disord. **43**, 34–44 (2012)
59. Perry, T.S.: Virtual reality goes social. IEEE Spectr. **53**, 56–57 (2016)
60. Riches, S., Elghany, S., Garety, P., Rus-Calafell, M., Valmaggia, L.: Factors affecting sense of presence in a virtual reality social environment: a qualitative study. Cyberpsychol. Behav. Soc. Netw. **22**, 288–292 (2019)
61. Singh, R.P., Javaid, M., Kataria, R., Tyagi, M., Haleem, A., Suman, R.: Significant applications of virtual reality for COVID-19 pandemic. Diabetes Metab. Syndr. Clin. Res. Rev. **14**, 661–664 (2020)
62. Ball, C., Huang, K.-T., Francis, J.: Virtual reality adoption during the COVID-19 pandemic: a uses and gratifications perspective. Telematics Inform. **65**, 101728 (2021)
63. Hobson, J.P., Williams, A.P.: Virtual reality: a new horizon for the tourism industry. J. Vacat. Mark. **1**, 124–135 (1995)
64. Zhang, S.-N., Li, Y.-Q., Ruan, W.-Q., Liu, C.-H.: Would you enjoy virtual travel? The characteristics and causes of virtual tourists' sentiment under the influence of the COVID-19 pandemic. Tour. Manag. **88**, 104429 (2021)
65. Elias, Z.M., Batumalai, U.M., Azmi, A.N.H.: Virtual reality games on accommodation and convergence. Appl. Ergon. **81**, 102879 (2019)
66. LaRocco, M.: Developing the 'best practices' of virtual reality design: Industry standards at the frontier of emerging media. J. Vis. Cult. **19**, 96–111 (2020)
67. Fernández, R.P., Alonso, V.: Virtual reality in a shipbuilding environment. Adv. Eng. Softw. **81**, 30–40 (2015)

68. Hirzle, T., Fischbach, F., Karlbauer, J., Jansen, P., Gugenheimer, J., Rukzio, E., Bulling, A.: Understanding, addressing, and analyzing digital eye strain in virtual reality head-mounted displays. ACM Trans. Comput. Interact. **29**, 1–80 (2022)
69. Laviola, J.J.: A discussion of cybersickness in virtual environments. ACM SIGCHI Bull. **32**, 47–56 (2000)
70. Kim, H.G., Lim, H.-T., Ro, Y.M.: Deep virtual reality image quality assessment with human perception guider for omnidirectional image. IEEE Trans. Circuits Syst. Video Technol. **30**, 917–928 (2019)
71. Stanney, K., Lawson, B.D., Rokers, B., Dennison, M., Fidopiastis, C., Stoffregen, T., Weech, S., Fulvio, J.M.: Identifying causes of and solutions for cybersickness in immersive technology: reformulation of a research and development agenda. Int. J. Hum. Comput. Interact. **36**, 1783–1803 (2020)
72. Desai, P.N.P.R., Desai, P.N.P.R., Ajmera, K.D., Mehta, K.: A review paper on Oculus Rift. Int. J. Eng. Trends Technol. **13**(4), 175–179 (2014). https://doi.org/10.14445/22315381/IJETT-V13P237
73. Kim, Y., Kim, H., Kim, Y.O.: Virtual reality and augmented reality in plastic surgery: a review. Arch. Plast. Surg. **44**(3), 179–187 (2017). https://doi.org/10.5999/aps.2017.44.3.179
74. Yung, R., Khoo-Lattimore, C.: New realities: a systematic literature review on virtual reality and augmented reality in tourism research. Curr. Issue Tour., 1–26 (2017). https://doi.org/10.1080/13683500.2017.1417359
75. Zhang, H.: Head-mounted display-based intuitive virtual reality training system for the mining industry. Int. J. Min. Sci. Technol. **27**(4), 717–722 (2017). https://doi.org/10.1016/j.ijmst.2017.05.005
76. Zhang, M., Zhang, Z., Chang, Y., Aziz, E.S., Esche, S., Chassapis, C.: Recent developments in game-based virtual reality educational laboratories using the Microsoft Kinect. Int. J. Emerg. Technol. Learn. (iJET). **13**, 138–159 (2018)
77. Shelstad, W.J., Smith, D.C., Chaparro, B.S.: Gaming on the rift: how virtual reality affects game user satisfaction. In: Proceedings of the Human Factors and Ergonomics Society 2017 Annual Meeting (2017). https://doi.org/10.1177/1541931213602001
78. Photo by Dana M. Barry
79. Allcoat, D., von Muhlenen, A.: Learning in virtual reality: effects on performance, emotion, and engagement. Res. Learn. Technol. **26**, 2140 (2018)
80. Lanzieri, N., McAlpin, E., Shilane, D., Samelson, H.: Virtual reality: an immersive tool for social work students to interact with community environments. Clin. Soc. Work. J. **49**(2), 207–219 (2021)
81. Hsu, Y.C.: Exploring the effectiveness of two types of virtual reality headsets for teaching high school mathematics. EURASIA J. Math. Sci. Technol. Educ. **17**(8), em1986 (2021)
82. Bos, D., Miller, S., Bull, E.: Using virtual reality (VR) for teaching and learning in geography: fieldwork, analytical skills, and employability. J. Geogr. High. Educ. **46**(3), 479–488 (2021). https://doi.org/10.1080/03098265.2021.1901867
83. Wiley, S.: File: Snowdon Ranger Path on a Cold February Day (16431627106).jpg (2015). File:Snowdon Ranger path on a cold February day. (16431627106).jpg—Wikimedia Commons

Chapter 9
Activities and Results for Individuals Engaging in Creative Lessons Using Virtual Reality Headsets

Dana M. Barry and Hideyuki Kanematsu

Abstract This chapter gives a general definition for a virtual reality (VR) headset, which can be used for VR video games, simulations, teaching purposes, and more. This headset is a head-mounted display that creates a reality-based virtual environment. It includes software that allows individuals to interact with real or imaginary computer-simulated environments. These three-dimensional experiences include both visual and auditory components. Our work presents and describes enjoyable educational activities, and the successful results of their creative lessons that were carried out by the authors and students while wearing VR headsets. For example, one person used the Oculus VR headset to experience parachuting from an airplane. He had an exciting adventure momentarily flying like a bird. Once the parachute opened, the individual observed the scenic landscape and quickly descended to the surface of the Earth with the aid of gravity. This VR activity could be used to provide exciting flight experiences, and creative lessons about flight, planes, sky jumping, birds, gravity, aeronautical engineering, and more. Some other thrilling lessons in this chapter involve the use of VR headsets to visit outer space, tour Jurassic Park, and have an exciting ride on a rollercoaster. Also included are college students in an introductory course about building virtual worlds for game design. They wore VR helmets to study virtual worlds (the background for video games) and to view and play video games to get ideas for creating their own video games (which include a story, characters, goals, and challenges). In addition, this chapter presents an activity where 20 fifth and sixth grade students used VR headsets to learn about ocean life in motion. They saw detailed, colorful, realistic images of life in the sea and felt they could almost touch the whales and sharks surrounding them. All the young

D. M. Barry (✉)
Clarkson University, Potsdam, NY, USA
e-mail: dbarry@clarkson.edu

H. Kanematsu
National Institute of Technology, Suzuka College, Suzuka, Mie, Japan
e-mail: kanemats@mse.suzuka-ct.ac.jp

D. M. Barry, H. Kanematsu (eds.), *Applications of Metaverse and Virtual Reality to Creative Education and Industry*, Intelligent Systems Reference Library 267, https://doi.org/10.1007/978-981-96-3341-8_9

participants (100%) said that learning with a VR headset is better than looking at 2-dimensional photos in a book.

Keywords Virtual reality (VR) headsets · Creative lessons using VR headsets · VR headsets provide realistic images · Studies show that students prefer VR headsets for learning

9.1 Introduction

This chapter begins by introducing the foundational aspects of virtual reality (VR) headsets [1–29]. A VR headset, often a head-mounted display, generates an immersive environment where individuals can interact with digitally simulated real or imaginary worlds. These three-dimensional (3-D) experiences incorporate both visual and auditory components, with the primary aim of creating an interactive space that stimulates the senses and transports users into a virtual reality/world. To enhance immersion, VR headsets are typically designed with a broad field of view to eliminate external distractions. These devices have been applied across various domains, including VR video games, simulations, and educational purposes.

Over time, Barry and Kanematsu conducted numerous effective problem-based learning (PBL) sessions in Second Life (SL), a 3-D virtual platform where avatars perform tasks on users' behalf [30]. PBL challenges students with complex, open-ended problems, guiding them through collaborative brainstorming sessions to develop solutions under instructors' mentorship. Recently, the authors expanded their interests to include VR headsets for teaching, while exploring their potential through literature research. They discovered studies that demonstrated how VR headsets could enhance students' academic performance and foster positive attitudes toward learning. Two illustrative studies are highlighted here.

The first study involved 99 first-year psychology students at the University of Warwick, UK [31]. Participants were randomly divided among three instructional methods: traditional textbooks, VR headsets, and 2-D video viewing. All groups engaged with the same educational content, including text and a 3-D plant cell model. The VR group, using HTC VIVE headsets, could interact with a dynamic 3-D cell model, enlarging parts and navigating within a virtual room where the cell appeared suspended. The video-based method presented a 2-D rendition of the same material, while the textbook group utilized static screenshots of the model. Pre- and post-learning assessments, along with self-rated emotional evaluations and questionnaires, revealed that the VR group excelled in memory retention and reported higher positive emotional engagement with reduced negative emotions compared to the other methods.

A second example involved a pilot project with social work students [28]. Using VR headsets, participants explored community environments via a 360-degree VR simulation. This immersive tool, designed to replicate a typical New York City neighborhood, helped students analyze how geographical factors and available resources influence communities. Participants utilized Google Daydream headsets with mobile devices to engage with panoramic urban settings. Data collection included pre- and post-surveys, as well as assessments of learning and attitudes. Findings showed notable improvements in post-test scores and positive feedback on the 360-degree simulation's utility, which participants described as engaging, motivating, and enjoyable.

Building on these insights, Barry and Kanematsu sought to leverage VR headsets to inspire students and enrich their learning experiences. To this end, they explored various VR devices through gaming and subject-specific video content, developing creative and interactive lesson plans. Some of these innovative educational activities are detailed in this chapter.

9.2 Playing Video Games to Generate Creative Ideas

Barry decided to experiment with virtual reality games first. Figure 9.1 shows her playing a game driving a car to earn points by avoiding as many obstacles as possible in a certain amount of time [32]. She appears happy and seems to be enjoying the game. In addition to entertainment, this game is useful for developing hand-eye coordination. Barry also played the video game King Kong in Skull Island while wearing a virtual reality headset. During this game, she drove a jeep around the island while being surrounded by many large monsters, snakes, dinosaurs, flying

Fig. 9.1 Barry avoids obstacles while driving a car to gain points during the game

Fig. 9.2 Barry prepares herself for a stressful adventure on Skull Island

reptiles, etc. She was tense, afraid, and screaming during the game. She tried to kill the monsters by waving her hands a certain way. Halfway through the game, another passenger joined her in the jeep and talked to her. He told her that the dinosaurs were angry, so he tried to help her kill them. This passenger looked and acted like a real person. However, he was a realistic part of the immersion experience. At the end, King Kong appears big and strong. This game could be used to obtain medical data. One could check a person's heartbeat, temperature, blood pressure, emotional status, etc. before, during, and after the game. Figure 9.2 shows Barry participating in a stressful VR game [32].

9.3 Immersive 3D Experiences That Inspire Innovative Lesson Ideas

Barry used an Oculus VR headset to experience an immersive journey through outer space. She found herself floating above the clouds, surrounded by the Sun and planets. The sensation was incredible, as if she were an astronaut drifting alone in the vastness of space. The experience brought her a sense of calm and relaxation. The planets appeared vibrant, highly detailed, and remarkably close.

In contrast, a similar experience in two dimensions involved viewing a colorful poster of the planets, which displayed their relative sizes and positions in space. While the poster provided an overview, it lacked the sense of presence and

Fig. 9.3 A photo of the planets is provided above

immersion offered by the VR experience. This virtual exploration of space can serve as a foundation for developing captivating lessons on topics such as space exploration, planetary science, lunar and Martian travel, aeronautical engineering, and careers in astronomy and astronautics. Figure 9.3 includes an image of the planets [33].

Kanematsu Used an Oculus VR Headset to Simulate a Parachuting Adventure He experienced the thrill of leaping from an airplane and briefly soaring through the sky like a bird. As the parachute deployed, he observed the breathtaking landscape below while gradually descending to the ground under the influence of gravity. In comparison, a similar two-dimensional experience involved viewing a photograph resembling the one shown in Fig. 9.4 [34]. This VR simulation can serve as the basis for engaging lessons on topics such as flight dynamics, aircraft, skydiving, avian flight, gravitational forces, aeronautical engineering, and related fields.

Barry Used a Google Cardboard VR Headset to Embark on a Thrilling Visit to Jurassic Park She experienced a mix of fear and excitement while exploring the

Fig. 9.4 This photo shows a man wearing a parachute as he prepares to land

park, which was filled with lush ferns and a variety of dinosaurs. She encountered gentle giants like the Brachiosaurus and fierce predators such as raptors and the Tyrannosaurus Rex. As she walked, the sound of distant roars echoed in the background, and she screamed when she came face-to-face with a Tyrannosaurus Rex. Fortunately, the creature ignored her and engaged in a battle with another T-Rex nearby. Barry was captivated by their movements, observing their massive limbs and sharp-toothed jaws during the intense fight. In contrast, a similar two-dimensional experience involved viewing a picture of various dinosaurs to compare their sizes and shapes. This VR journey could be utilized for immersive tours of Jurassic Park, lessons about dinosaurs, or studies on fascinating time periods like the Jurassic Period, Mesozoic Era, and Triassic Period.

9.4 Exploring Rollercoasters Through Virtual Reality Headsets

Ten-year-old students at Trinity Catholic School in Massena, New York, used Google Cardboard VR headsets to explore the exciting world of rollercoasters and simulate the experience of riding one. Each of the ten participants watched the same video about rollercoasters, which served as a control for the project. Figure 9.5 displays a rollercoaster like the one depicted in the virtual reality experience [35]. This activity allowed students to learn about various aspects of rollercoasters, including their design, dimensions, colors, and structures. Additionally, the VR experience provided them with a lifelike simulation of a rollercoaster ride. To gather feedback, the students completed a survey designed to evaluate their virtual experience.

Survey Form
Circle your answers to the following questions.

1. Were you impressed with the virtual reality experience?

   (a) very much, (b) pretty much, (c) neutral, (d) not so much, (e) not at all

Fig. 9.5 This image shows a rollercoaster like the one experienced during the virtual reality activity

2. Did you enjoy the experience?

   (a) very much, (b) pretty much, (c) neutral, (d) not so much, (e) not at all

3. Were you motivated to learn more?

   (a) very much, (b) pretty much, (c) neutral, (d) not so much, (e) not at all

4. Do you feel this type of learning style is better than looking at a two-dimensional photo?

   (a) very much, (b) pretty much, (c) neutral, (d) not so much, (e) not at all

Completion questions

5. What did you like best about the virtual reality experience?

   ____

6. What other topic would you like to experience using virtual reality headsets?

   ____

The survey results indicate that 90% of participants reported being either "pretty much" or "very much" impressed by the experience, and 100% stated that they "pretty much" to "very much" enjoyed it. Regarding preferred learning styles, 90% of the students favored the virtual reality approach. Additionally, 80% expressed an interest in exploring more VR-based learning activities.

Students' Comments
Many participants mentioned that they thoroughly enjoyed the ride, describing it as closely resembling the experience of being on a real rollercoaster. Some noted that they particularly liked the view during the ride. Suggestions for future VR experiences included activities such as virtual travel to different locations, riding water park slides, experiencing thrilling chases, encountering scary situations, and engaging in science projects.

9.5 Exploring Ocean Life Through Virtual Reality Headsets

Twenty fifth- and sixth-grade students at Trinity School in Massena, New York, used Google Cardboard VR headsets to explore the dynamic world of ocean life. While soft music played in the background, the students immersed themselves in a 3-D video showcasing marine creatures gliding through the water. To minimize variables in the project, all students viewed the same video, which acted as a control.

This activity provided an engaging educational experience by presenting vibrant, detailed, and lifelike depictions of aquatic animals and their movements. One notable example is the gray reef shark depicted in Fig. 9.6 [36]. This shark exhibits a sleek, streamlined body and a side-to-side swimming motion resembling a figure eight. Known as one of the fastest and rarest species, it inhabits the Pacific and

Fig. 9.6 This photo shows a grey reef shark

Indian Oceans. Figure 9.7 displays a Google Cardboard VR headset [37], which allows users to view 3-D videos on platforms like YouTube using an iPhone.

To gather further insights into the experience, students completed a survey form. Their responses to questions 1–6 are represented in the bar graphs provided in Figs. 9.8, 9.9, 9.10, 9.11, 9.12, and 9.13. Note: The blue columns indicate the responses from our students.

Survey Form
Circle your answers to the following questions.

1. Were you impressed with the virtual reality experience?

   (a) very much, (b) pretty much, (c) neutral, (d) not so much, (e) not at all

2. Did you enjoy the virtual reality experience?

   (a) very much, (b) pretty much, (c) neutral, (d) not so much, (e) not at all

3. Did the VR experience make you feel relaxed?

   (a) very much, (b) pretty much, (c) neutral, (d) not so much, (e) not at all

4. Did the virtual reality headset motivate you to learn more than a 2-D photo would?

   (a) very much, (b) pretty much, (c) neutral, (d) not so much, (e) not at all

5. Did the virtual reality headset cause you to be more focused (less distracted from the topic) than looking at 2-D photos?

   (a) very much, (b) pretty much, (c) neutral, (d) not so much, (e) not at all

Fig. 9.7 This is a photo of a Google cardboard virtual reality headset

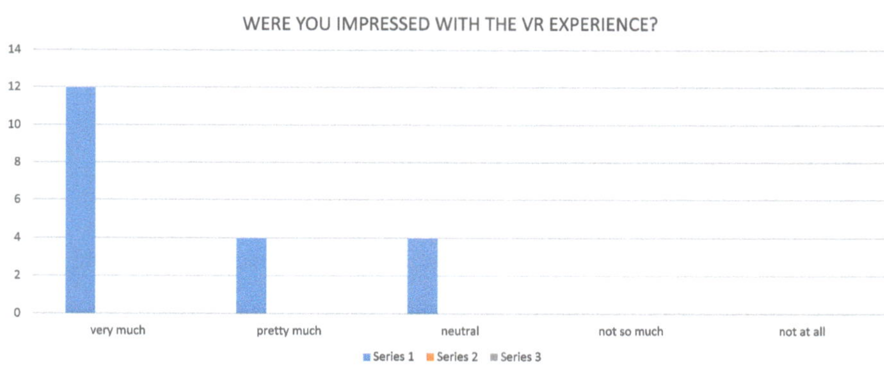

Fig. 9.8 Bar graph for question number 1 on the survey form

6. Do you feel that learning with a virtual reality headset is better than looking at two-dimensional photos in a book?

   (a) very much, (b) pretty much, (c) neutral, (d) not so much, (e) not at all

Completion questions

7. What did you like best about the virtual reality experience?

   ____

Fig. 9.9 Bar graph for question number 2 on the survey form

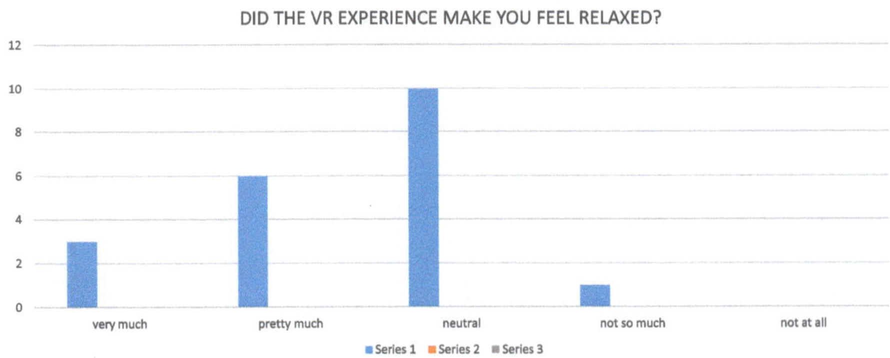

Fig. 9.10 Bar graph for question number 3 on the survey form

8. What other topics would you like to experience using virtual reality headsets?

   ____

9. Other comments:

   ____

   ____

9.6 Data Analysis for the Activity About Ocean Life in Motion Using VR Headsets

The Answers to the Short Survey Questions (1–6) Provide the Following Information Most of the participants (90%) enjoyed the virtual reality experience and 90% of them said that using the virtual reality headset caused them to be more focused (less distracted from the topic) than looking at a two-dimensional photo. All

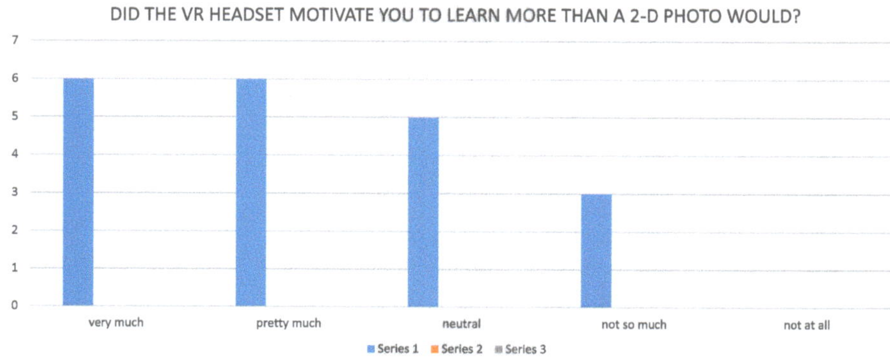

Fig. 9.11 Bar graph for question number 4 on the survey form

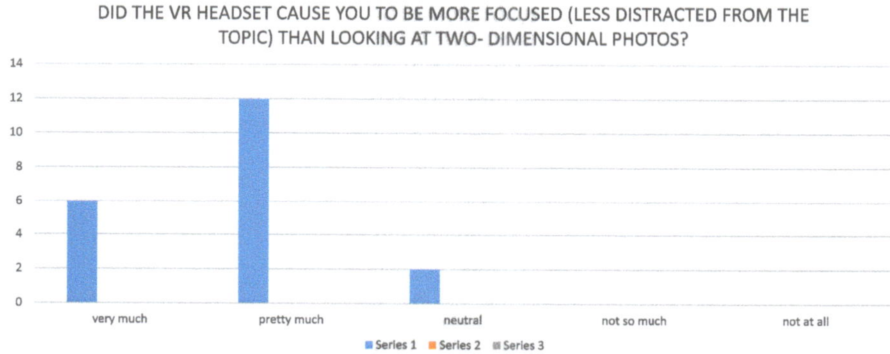

Fig. 9.12 Bar graph for question number 5 on the survey form

students (100%) felt that learning with a virtual reality headset is better than looking at two-dimensional photos in a book. On the other hand, half of the participants (50%) had a neutral feeling about being relaxed during the virtual reality experience.

Question Number 7 This question asked the participants what they liked best about the VR experience. Many said they felt like they were really in the ocean and almost able to touch the sharks, etc. in that 3-dimensional environment. Some liked the ocean along with the relaxing sounds and music. Others liked viewing the sea life in motion, the corals, the colorful images, and more. Most of the students mentioned that the experience was more educational and entertaining than looking at two-dimensional photos.

Question Number 8 This question asked participants about other topics they would like to explore using VR headsets. Their responses included suggestions such as exploring the jungle, observing tropical animals and plants, exploring the woods, engaging in math and other school subjects, experiencing rollercoasters, participating in escape room scenarios, viewing waterfalls, exploring sports and engineering

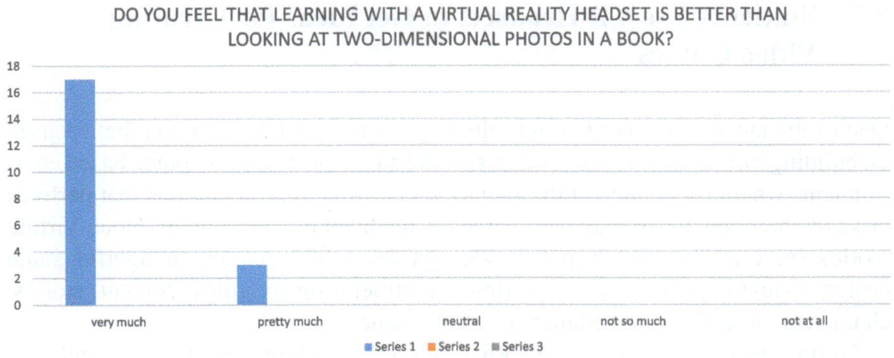

Fig. 9.13 Bar graph for question number 6 on the survey form

concepts, observing nature, visiting various global locations, studying animals beyond marine life, learning about dinosaurs, exploring farming, and simulating the experience of flying like a bird.

Question Number 9 This section recorded additional comments. Some students described the 3-dimensional experience as entertaining, impressive, realistic, and a more effective teaching tool compared to 2-dimensional methods. Others expressed interest in using VR headsets for educational purposes in their classrooms, believing this would make lessons more engaging and help students focus better. They noted that the headset's wide field of view and reduced external distractions contributed to its effectiveness. However, a few students reported that the 3-dimensional video appeared blurry, possibly due to vision issues or improper adjustment of the headset's focus settings. Additionally, one student mentioned experiencing dizziness during the session, a condition known as cyber sickness, which is like motion sickness [26].

Overall Comment The VR activity about sea life in motion was very successful. The students enjoyed it very much and obtained lots of information about sea life (like color, size, shape, and type of movements in water of an octopus, sharks, sting rays, whales, and others. They obtained this data by using VR headsets to view a realistic 3-dimensional video about sea life. In addition, all the participants (100%) felt that learning with a VR headset is better than looking at 2-dimensional photos in a book.

9.7 Students Use VR Headsets to Get Ideas for Creating Video Games

One of the authors observed an introductory course at SUNY Canton that focused on building virtual worlds for game design. Virtual worlds are computer-based environments where users interact through avatars (virtual representations that perform tasks on their behalf), or playable characters included in video games. Note: Virtual worlds serve as the backdrop for video games, which require thoughtful game design. Game design involves creating and structuring the rules, content, stories, characters, objectives, and challenges of the game.

During the class, students wore VR headsets to explore virtual worlds and play video games, gathering inspiration for designing their own games. They analyzed what aspects were enjoyable and easy versus those that were frustrating or tedious. Students were introduced to different types of games, ranging from those for children to those for adults. Action games, which emphasize physical challenges like combat requiring hand-eye coordination, are particularly popular. Adventure and survival games were also highlighted as intriguing options. Each student selected a specific game type to design, which required building a virtual world and creating a game to complement it.

It's worth noting that while anyone can join a virtual world platform like Second Life (SL) to create their own space, the college students were tasked with building virtual worlds from scratch. This process requires a game engine and a modeling package. Typically, this includes tools such as Autodesk Maya (a 3D computer graphics application for animating characters), Substance Painter (for coloring and texturing), and Unreal Engine (a comprehensive game engine with various creation tools).

The main objective of video games is immersion, which is significantly enhanced by VR helmets. These devices provide a broad field of vision, minimizing external distractions. Video games can range from realistic to cartoonish, abstract, or artistic. Although creating a video game from scratch is time-consuming, the students successfully completed the required stages of their projects. They enjoyed the immersive VR experiences and expressed excitement about their virtual worlds and game designs. Some even envisioned developing successful games for future marketing.

9.8 Conclusions

This chapter gives a general definition for a virtual reality (VR) headset, which can be used for VR video games, simulations, teaching purposes, and more. This headset is a head-mounted display that creates a reality-based virtual environment. It includes software that allows individuals to interact with real or imaginary computer-simulated environments. These three-dimensional experiences include both visual and auditory components. Our work presents and describes enjoyable educational activities, and the successful results of their creative lessons that were carried out by the authors and students while wearing VR headsets. For example, one person used the Oculus VR headset to experience parachuting from an airplane. He had an exciting adventure momentarily flying like a bird. Once the parachute opened, the individual observed the scenic landscape and quickly descended to the surface of the Earth with the aid of gravity. This VR activity could be used to provide exciting flight experiences, and creative lessons about flight, planes, sky jumping, birds, gravity, aeronautical engineering, and more. Some other thrilling lessons in this chapter involve the use of VR headsets to visit outer space, tour Jurassic Park, and have an exciting ride on a rollercoaster. Also included are college students in an introductory course about building virtual worlds for game design. They wore VR helmets to study virtual worlds (the background for video games) and to view and play video games to get ideas for creating their own video games (which include a story, characters, goals, and challenges). In addition, this chapter presents an activity where 20 fifth and sixth grade students used VR headsets to learn about ocean life in motion. They saw detailed, colorful, realistic images of life in the sea and felt they could almost touch the whales and sharks surrounding them. All the fifth and sixth grade students (100%) said that learning with a VR headset is better than looking at 2-dimensional photos in a book. A concluding statement for this chapter is that the participants who wore VR helmets for our creative lessons gained lots of information about various topics and thoroughly enjoyed being immersed in a variety of 3-dimensional virtual reality environments.

References

1. Al Awadhi, et al.: Virtual reality application for interactive and informative learning. In: 2nd International Conference on Bioengineering for Smart Technologies, pp. 1–4. IEEE (2017)
2. Biocca, F., Delaney, B.: Immersive Virtual Reality Technology Communication in the Age of Virtual Reality, pp. 57–124. Lawrence Erlbaum Associates, Inc, Hillsdale (1995)
3. Bowman, D.A., McMahan, R.P.: Virtual reality: how much immersion is enough? Computer. **40**(7), 36–43 (2007)
4. Carruth, D.W.: Virtual Reality for Education and Workforce Training 15th International Conference on Emerging eLearning Technologies and Applications, pp. 1–6. IEEE (2017)
5. Chavez, B., Bayona, S.: Virtual Reality in the Learning Process: Trends and Advances in Information Systems and Technologies, pp. 1345–1356. Springer (2018)
6. Chin, N., Gupte, A., Nguyen, J., Sukhin, S., Wang, G., Mirizio, J.: Using virtual reality for an immersive experience in the water cycle. In: IEEE MIT Undergraduate Research Technology Conference, pp. 1–4. IEEE (2017)
7. Dolezal, M., Chmelik, J., Liarokapis, F.: An Immersive Virtual Environment for Collaborative Geo-Visualization, pp. 272–275. VS-GAMES (2017)
8. Farra, S.L., Smith, S.J., Ulrich, D.L.: The student experience with varying immersion levels of virtual reality simulation. Nurs. Educ. Perspect. **39**(2), 99–101 (2018)
9. Feng, Z., González, V.A., Amor, R., Lovreglio, R., Cabrera-Guerrero, G.: Immersive virtual reality serious games for evacuation training and research: a systematic literature review. Comput. Educ. **127**, 252–266 (2018)

10. Gerloni, I.G., Carchiolo, V., Vitello, F.R., Sciacca, E., Becciani, U., Costa, A., et al.: Immersive virtual reality for earth sciences federated conference on computer science and information system. IEEE, pp. 527–534 (2018)
11. Jensen, L., Konradsen, F.: A review of the use of virtual reality head-mounted displays in education and training. Educ. Inf. Technol. **23**(4), 1515–1529 (2018). https://doi.org/10.1007/s10639-017-9676-0
12. Kwon, B., Kim, J., Lee, K., Lee, Y.K., Park, S., Lee, S.: Implementation of a virtual training simulator based on 360° multi-view human action recognition. IEEE Access. **5**, 12496–12511 (2017). https://doi.org/10.1109/ACCESS.2017.2723039
13. Martín-Gutiérrez, J., Mora, C.E., Añorbe, D.B., González-Marrero, A.: Virtual technologies trends in education. EURASIA J. Math. Sci. Technol. Educ. **13**(2), 469–486 (2017)
14. Merchant, Z., Goetz, E.T., Cifuentes, L., Keeney-Kennicutt, W., Davis, T.J.: Effectiveness of virtual reality-based instruction on students' learning outcomes in k-12 and higher education: a meta-analysis. Comput. Educ. **70**, 29–40 (2014)
15. Papachristos, N.M., Vrellis, I., Mikropoulos, T.A.: A comparison between Oculus rift and a low-cost smartphone VR headset: immersive user experience and learning. In: IEEE 17th International Conference on Advanced Learning Technologies, pp. 477–481 (2017). https://doi.org/10.1109/ICALT.2017.145
16. Parong, J., Mayer, R.E.: Learning science in immersive virtual reality. J. Educ. Psychol. **110**(6), 785–797 (2018)
17. Suh, A., Prophet, J.: The state of immersive technology research: a literature analysis. Comput. Hum. Behav. **86**, 77–90 (2018)
18. Ye, Q., Hu, W., Zhou, H., Lei, Z., Guan, S.: VR interactive feature of html5-based webvr control laboratory by using head-mounted displays. Int. J. Online Eng. **14**(3) (2018)
19. Velev, D., Zlateva, P.: Virtual reality challenges in education and training. Int. J. Learn. Teach. **3**, 33–37 (2017)
20. Radianti, J., Majchrzak, T.A., Fromm, J., Wohlgenannt, I.: A systematic review of immersive virtual reality applications for higher education: design elements, lessons learned, and research agenda. Comput. Educ. **147**, 103778 (2020). https://doi.org/10.1016/j.compedu.2019.103778
21. Lau, K.W., Lee, P.Y.: The use of virtual reality for creating unusual environmental stimulation to motivate students to explore creative ideas. Interact. Learn. Environ. **23**, 3–18 (2012)
22. Hamilton, D., McKechnie, J., Edgerton, E., Wilson, C.: Immersive virtual reality as a pedagogical tool in education: a systematic literature review of quantitative learning outcomes and experimental design. J. Comput. Educ. **8**, 1–32 (2020)
23. Morozov, M., Gerasimov, A., Fominykh, M., Smorkalov, A.: Transactions on Computational Science XVIII Asynchronous immersive classes in a 3D virtual world: extended description of v Academia, vol. 7848, pp. 81–100. Springer, Cham (2013)
24. Bird, J.M.: The use of virtual reality head-mounted displays within applied sport psychology. J. Sport Psychol. Action. **11**, 115–128 (2020)
25. Hirzle, T., Fischbach, F., Karlbauer, J., Jansen, P., Gugenheimer, J., Rukzio, E., Bulling, A.: Understanding, addressing, and analyzing digital eye strain in virtual reality head-mounted displays. ACM Trans. Comput. Interact. **29**, 1–80 (2022)
26. Laviola, J.J.: A discussion of cybersickness in virtual environments. ACM SIGCHI Bull. **32**, 47–56 (2000)
27. Zhang, H.: Head-mounted display-based intuitive virtual reality training system for the mining industry. Int. J. Min. Sci. Technol. **27**(4), 717–722 (2017). https://doi.org/10.1016/j.ijmst.2017.05.005
28. Lanzieri, N., McAlpin, E., Shilane, D., Samelson, H.: Virtual reality: an immersive tool for social work students to interact with community environments. Clin. Soc. Work. J. **49**(2), 207–219 (2021)
29. Hsu, Y.C.: Exploring the effectiveness of two types of virtual reality headsets for teaching high school mathematics. EURASIA J. Math. Sci. Technol. Educ. **17**(8), em1986 (2021)

30. Barry, D.M., Kanematsu, H., Fukumura, Y., Kobayashi, T., Ogawa, N., Nagai, H.: Problem-based learning activities in second life. Int. J. Modern Educ. Forum (IJMEF). **3**(1) (2014). www.ijmef.org
31. Allcoat, D., von Muhlenen, A.: Learning in virtual reality: effects on performance, emotion, and engagement. Res. Learn. Technol. **26**, 2140 (2018)
32. Photo by Eric Barry
33. Crane, C.S.: File: Planet collage to scale.jpg. License: Creative Commons Attribution-Share Alike 4.0 international license (2022). File:Planet collage to scale.jpg—Wikimedia Commons
34. Ringstone, A.: File: Army display parachutist landing 29Sept2018arp.jpg. License: This work is in the public domain (2018). File:Army display parachutist landing 29Sept2018 arp.jpg—Wikimedia Commons
35. Hmich 176 at English Wikipedia.: File: Kingdom Coaster 002.JPG. License: This license is under the Creative Commons Attribution- ShareAlike 3.0 unported license. (Attribution: Hmich 176 at English Wikipedia) (2011). File:Kingdom Coaster 002.JPG—Wikimedia Commons
36. NOAA Photo Library.: File: Corlo207 (28225976491).jpg. License: Creative Commons Attribution 2.0 Generic license (2015). File:Corl0207 (28225976491).jpg—Wikimedia Commons
37. Photo by Dana Barry

Chapter 10
Physiological Results Obtained from Individuals Using Virtual Reality (VR) Headsets Along with Google Sensor Glasses

Hideyuki Kanematsu and Dana M. Barry

Abstract This chapter explores the integration of eye potential sensors with Virtual Reality (VR) headsets to capture detailed physiological data, providing insights into cognitive and emotional responses within immersive environments. Our research involved the use of advanced sensor glasses in conjunction with the Meta Quest 2 VR headset to monitor eye movements, blinking patterns, and other physiological indicators. Through a series of experiments, we demonstrated how these physiological metrics can inform the design of VR applications, enhance user experience, and contribute to the development of e-learning platforms. Additionally, we delve into the 'fight or flight' response theory to underscore the importance of physiological measurements in understanding psychological states. This chapter highlights the potential of combining VR with real-time physiological data to create adaptive, personalized learning experiences that can overcome traditional educational barriers.

Keywords Virtual reality (VR) · Physiological measurements · Eye potential sensors · Fight-or-flight response · E-learning

10.1 Introduction

In recent years, the intersection of virtual reality (VR) technology and physiological monitoring has emerged as a frontier of research, offering profound implications for understanding human interaction with digital environments [1–3]. This chapter

H. Kanematsu (✉)
National Institute of Technology, Suzuka College, Suzuka-shi, Mie, Japan
e-mail: hideyuki.kanematsu@bioenglab.org

D. M. Barry
Clarkson University, Potsdam, NY, USA
e-mail: dbarry@clarkson.edu

D. M. Barry, H. Kanematsu (eds.), *Applications of Metaverse and Virtual Reality to Creative Education and Industry*, Intelligent Systems Reference Library 267, https://doi.org/10.1007/978-981-96-3341-8_10

delves into the innovative integration of sensor glasses with VR headsets, a technological synergy that paves the way for capturing intricate physiological data within immersive virtual landscapes. Our exploration is anchored in a detailed examination of a specialized VR headset and sensor glasses designed to harness physiological signals, thereby unlocking new dimensions of user experience and interaction analysis.

The proliferation of VR technology, exemplified by devices such as the Meta Quest 2, has revolutionized the way individuals engage with digital content, blurring the lines between physical and virtual realms [4–6]. The sensation of presence, or the feeling of being "inside" a virtual space, is a cornerstone of effective VR experiences. This chapter introduces an experiment that leverages the Meta Quest 2 headset to transport users to a meticulously rendered 3D simulation of a hot spring hotel situated at the picturesque base of a volcano. This setting provides a visually captivating backdrop and a unique context for studying physiological responses within a VR environment.

Central to our investigation is the employment of sensor glasses developed by JINS MEME, which utilize the eye's natural potential differences to gather data on blinking patterns. This innovative approach is based on the understanding that the eye exhibits a positive potential on the cornea side and a negative potential on the fundus side, with movements such as blinking altering this potential distribution. Such changes are indicative of various cognitive states, including concentration, vitality, and calmness, offering a window into the user's engagement and emotional state during the VR experience.

Furthermore, the incorporation of an ophthalmic sensor capable of tracking head movements complements the eye-potential data, enriching our analysis of user interaction within the virtual space. This dual-sensor approach facilitates a comprehensive understanding of how participants navigate and respond to VR environments, shedding light on the intricate relationship between physical actions and psychological states.

This chapter aims to contribute to the burgeoning field of VR research by providing an in-depth look at the methods and technologies employed to capture physiological data in virtual environments. Through the lens of a specific experiment and a review of related studies, we seek to illuminate the potential of combining VR with advanced sensing technologies for enhanced user experience research. Our discussion extends beyond the technical implementation to consider the implications of these findings for the design and development of VR applications, highlighting the pivotal role of physiological data in crafting immersive and responsive virtual worlds.

10.2 What Is Virtual Reality (VR)?

Virtual Reality (VR) is an immersive technology that allows users to interact with computer-generated environments in a way that simulates a real-world or imagined experience. Unlike traditional interfaces, where users view a screen, VR places the user inside an experience, allowing them to engage with three-dimensional (3D) worlds. This is typically achieved through the use of VR headsets, which encompass displays, motion tracking, and other sensory inputs to create a convincing sense of presence, or the feeling of being physically present in the non-physical world.

10.2.1 Evolution and Development of VR

The concept of VR is not new; it has its roots in the 1960s with early pioneers like Morton Heilig, who created the Sensorama, a multimedia device that delivered visual, auditory, and tactile sensations to simulate a real-world experience. However, it wasn't until the development of computer technology that VR began to take its modern form. The 1980s saw significant advancements with the introduction of head-mounted displays (HMDs) by companies like VPL Research, founded by Jaron Lanier, who is often credited with popularizing the term "virtual reality" [7].

In the 1990s, VR technology began to enter the public consciousness. However, the technology was still in its infancy and primarily used in specialized applications such as military training, flight simulation, and high-end scientific research. The prohibitive cost and technical limitations, such as low-resolution displays and significant latency, meant that the general public needed to adopt VR widely.

The twenty-first century revolutionized VR technology, largely driven by advancements in computer graphics, display technology, and motion tracking. The release of the Oculus Rift in 2012 marked a turning point, demonstrating that high-quality VR experiences could be delivered at a consumer-friendly price point. Since then, companies like HTC, Sony, and Meta (formerly Facebook) have continued to push the boundaries of VR technology, making it more accessible and widely adopted across various industries [8–10].

10.2.2 Core Components of VR Systems

A typical VR system consists of several key components:

1. **Head-Mounted Display (HMD):** The HMD is the most recognizable part of a VR system, housing the displays and lenses that provide stereoscopic 3D visuals. Modern HMDs also include sensors for tracking the user's head movements, allowing the view in the virtual world to change in real-time as the user looks

around. Some advanced HMDs also include built-in headphones for spatial audio, further enhancing the immersive experience.
2. **Motion Tracking:** VR systems use various technologies to track the user's movements, including head tracking, hand tracking, and body tracking. Head tracking is typically achieved using a combination of gyroscopes, accelerometers, and magnetometers, while hand and body tracking may use external cameras, sensors, or controllers equipped with tracking markers. This allows the system to accurately replicate the user's movements within the virtual environment, creating a more natural and intuitive experience.
3. **Input Devices:** Interacting with the virtual environment requires specialized input devices, such as handheld controllers, gloves, or even full-body suits equipped with sensors. These devices allow users to manipulate objects, navigate through virtual spaces, and perform a wide range of actions within the VR world.
4. **Computing Hardware:** Rendering a convincing VR experience requires significant computing power. The system must generate high-resolution 3D graphics at a smooth frame rate while processing user inputs and updating the environment in real-time. This is typically handled by a powerful PC or, in the case of standalone VR headsets, an integrated processing unit.
5. **Software:** The software component of a VR system includes the virtual environments, simulations, or games that users interact with. These programs are designed to take full advantage of the hardware's capabilities, delivering experiences that range from simple 3D visualizations to complex, interactive worlds.

10.2.3 Applications of VR

VR has found applications across a wide range of fields, transforming industries and creating new possibilities for human interaction and experience.

1. **Gaming:** Gaming remains one of the most popular applications of VR, with titles like "Beat Saber," "Half-Life: Alyx," and "Resident Evil 4 VR" showcasing the potential of the medium. VR allows gamers to immerse themselves in their favorite worlds, interacting with the environment and characters in ways that traditional gaming cannot match [11].
2. **Education and Training:** VR offers unique opportunities for education and training by providing safe, controlled environments where users can practice skills or explore new concepts. Medical students, for example, can perform virtual surgeries to hone their skills without the risk of harming a patient, while astronauts can train for space missions in a simulated zero-gravity environment [12].
3. **Healthcare:** Beyond training, VR is also being used for therapeutic purposes, such as exposure therapy for phobias, pain management, and rehabilitation.

VR's ability to create controlled, immersive environments makes it a powerful tool for mental and physical health treatments [13].

4. **Architecture and Design:** VR allows architects and designers to visualize and explore their creations in a fully immersive environment, making it easier to identify potential issues and make adjustments before construction begins. Clients can also experience the design in VR, providing a better understanding of the space than traditional blueprints or 2D renderings [14].
5. **Social Interaction:** VR has the potential to transform the way we interact with others, enabling virtual meetings, social gatherings, and collaborative workspaces. Platforms like VRChat and AltspaceVR allow users to connect with people from around the world in a shared virtual space, complete with avatars and spatial audio [15].
6. **Tourism and Exploration:** VR can transport users to far-off places, allowing them to explore new destinations from the comfort of their own home. Virtual tours of landmarks, museums, and natural wonders provide an accessible way to experience the world, even for those who may not be able to travel physically [16].

10.2.4 The Future of VR

As VR technology continues to evolve, its potential applications are likely to expand even further. Advances in display technology, such as higher resolution and wider fields of view, will make VR experiences more immersive and realistic. Improvements in motion tracking and input devices will make interactions within virtual environments more natural and intuitive. Additionally, the integration of artificial intelligence and machine learning could enable more responsive and personalized VR experiences, adapting to the user's preferences and behaviors in real-time.

In conclusion, VR represents a significant shift in how we interact with digital content, offering immersive experiences that were previously unimaginable. As technology matures, it will continue to reshape industries, create new opportunities for innovation, and change how we experience the world.

10.3 The Importance of Physiological Measurements in Analyzing Psychological States—The Fight or Flight Response as a Foundation

Physiological measurements are a cornerstone in the study of human psychology and behavior, offering a direct window into the body's response to various stimuli. Among the many theories that bridge the gap between psychology and physiology, the 'fight or flight' response stands out as a foundational concept. First introduced by American physiologist Walter B. Cannon in the early twentieth century, the 'fight or flight' response describes the body's automatic reaction to perceived threats, involving the rapid activation of the sympathetic nervous system. This response is characterized by a series of physiological changes, including increased heart rate, accelerated breathing, and heightened blood flow to muscles, all of which prepare the body to either confront the danger (fight) or escape from it (flight) [17].

10.3.1 The Role of the Fight or Flight Response in Understanding Psychological States

The 'fight or flight' response is more than just a physiological reaction; it is a critical mechanism for understanding the psychological state of an individual. When faced with a stressful situation, the body's immediate physical responses are directly tied to the individual's mental state, providing tangible indicators of fear, anxiety, stress, or excitement. This connection between mind and body is why physiological measurements are so valuable in psychological research. By measuring physical reactions such as heart rate, respiration, and perspiration, researchers can gain insights into the underlying psychological states that drive these responses.

For example, an increased heart rate can indicate not only physical exertion but also emotional arousal, whether due to fear, excitement, or stress. Similarly, changes in skin conductance (often measured as galvanic skin response) reflect the activity of sweat glands, which are controlled by the sympathetic nervous system. This can serve as a proxy for emotional arousal, making it possible to infer psychological states from physiological data.

10.3.2 Application of the Fight or Flight Theory in VR Research

In the context of Virtual Reality (VR), understanding the 'fight or flight' response becomes particularly relevant. VR environments can be designed to simulate a wide range of scenarios, from peaceful landscapes to high-stress situations that might trigger a 'fight or flight' response. By measuring physiological indicators during VR experiences, researchers can assess how different virtual scenarios affect users psychologically.

For instance, in a VR scenario where a user is placed in a dangerous or threatening situation, physiological measurements such as heart rate, breathing rate, and skin conductance can provide real-time data on the user's psychological state. If these measurements show an increase in sympathetic nervous system activity, it suggests that the user is experiencing a 'fight or flight' response, indicating high levels of stress or fear. This data is invaluable for understanding how users interact with VR environments and can inform the design of more effective and immersive experiences.

10.3.3 Measuring Heart Rate and Its Psychological Implications

Heart rate is one of the most straightforward and widely used physiological measurements in psychological research [18]. When the body encounters a stressor, the heart rate typically increases as part of the 'fight or flight' response. This increase is driven by the release of adrenaline, which prepares the body for quick action. In a controlled VR environment, monitoring heart rate can help researchers determine which aspects of the virtual experience are most likely to induce stress or arousal.

For example, in a study where participants navigate a virtual maze filled with potential threats, heart rate data can reveal which moments or elements of the maze cause the most stress. A sudden spike in heart rate when encountering a virtual predator, for example, would indicate a strong psychological response, likely tied to the user's perception of danger. This kind of data can be used to refine VR environments to either amplify or mitigate stress, depending on the desired outcome of the research or application.

10.3.4 The Significance of Respiration Rate in Psychological Analysis

Respiration rate is another critical physiological measure linked to the 'fight or flight' response [19]. When an individual perceives a threat, their breathing rate increases to supply more oxygen to the muscles, preparing the body for rapid action. In VR research, monitoring respiration rate provides insights into how immersive environments affect a user's psychological state.

For example, during a VR experience designed to simulate a high-stress situation, such as a firefighting scenario, researchers can measure changes in respiration rate to assess the level of stress experienced by the user. A rapid increase in breathing rate might indicate that the user is highly engaged and possibly stressed by the scenario, reflecting a heightened 'fight or flight' response. This data can be crucial for training applications, where understanding the physiological impact of stress on performance is essential.

10.3.5 Skin Conductance as a Measure of Emotional Arousal

Skin conductance, or galvanic skin response (GSR), is a direct measure of sweat gland activity, which is closely tied to emotional arousal [20]. When the sympathetic nervous system is activated, as in the 'fight or flight' response, sweat production increases, leading to higher skin conductance. In psychological research, GSR is often used to measure emotional responses to stimuli, providing a non-invasive way to assess how individuals react to different situations.

In VR research, GSR can be particularly useful for understanding how different elements of a virtual environment influence emotional arousal. For instance, in a VR horror game, researchers might measure GSR to determine which parts of the game are most effective at inducing fear. A sharp increase in skin conductance when encountering a frightening scene would suggest a strong emotional response, indicating that the user is experiencing a 'fight or flight' reaction.

10.3.6 The Role of Facial Expressions and Blinking in Psychological Analysis

Facial expressions and blinking patterns are also valuable indicators of psychological states [21–25]. While these are not directly part of the 'fight or flight' response, they can provide complementary data that helps to paint a more complete picture of a user's mental state. For example, frequent blinking may indicate cognitive load or stress, while facial expressions can reveal emotions such as fear, surprise, or happiness.

In VR research, tracking facial expressions and blinking can provide additional context to the physiological data collected from heart rate, respiration, and skin conductance measurements. For example, if a user exhibits a fearful facial expression and increased blinking during a VR experience, alongside elevated heart rate and GSR, it strongly suggests that the user is experiencing a 'fight or flight' response.

10.3.7 Integrating Physiological Measurements into VR-Based Psychological Research

The integration of physiological measurements into VR-based psychological research allows for a more nuanced understanding of how virtual environments affect users. By measuring the body's physiological responses, researchers can gain insights into the underlying psychological processes that drive behavior. This data is not only valuable for research purposes but also has practical applications in fields such as education, training, and therapy.

For example, in educational VR environments, understanding how students respond physiologically to different types of content can help educators design more effective learning experiences. If physiological data indicates that certain content is causing excessive stress, it might be beneficial to adjust the material to reduce cognitive load and improve learning outcomes. Conversely, if the data shows that students are not sufficiently engaged, more stimulating content can be introduced to capture their attention.

In therapeutic settings, VR combined with physiological measurements can be used to help individuals confront and manage stress or anxiety. By monitoring physiological responses in real-time, therapists can adjust the virtual environment to create a safe and controlled space for exposure therapy, helping patients gradually overcome their fears.

10.3.8 Conclusion

The 'fight or flight' response theory provides a powerful framework for understanding the relationship between physiological measurements and psychological states. By leveraging this theory in the context of VR research, we can gain valuable insights into how users react to virtual environments, both mentally and physically. Physiological measurements such as heart rate, respiration rate, skin conductance, and facial expressions offer a direct link to the autonomic nervous system's activity, providing real-time data on a user's psychological state. This approach not only enhances our understanding of human behavior in digital environments but also has practical applications in education, training, and therapy, where it can be used to create more personalized and effective experiences. As VR technology continues to evolve, the integration of physiological measurements will play an increasingly important role in shaping the future of psychological research and digital interaction.

10.4 Physiological Measurements Using Apparatuses

Physiological measurements provide valuable insights into how individuals respond to various stimuli, including those presented in virtual reality (VR) environments. These measurements allow researchers to objectively assess aspects of a person's physical and psychological state while they interact with virtual worlds. In this section, we will discuss the general approaches to physiological measurements and specifically focus on the measurement of electroencephalography (EEG), pulse rate, perspiration rate, body temperature, facial expressions, and blinking.

10.4.1 Overview of Physiological Measurements

Physiological measurements are typically non-invasive methods used to monitor and record biological signals that indicate the functioning of various bodily systems. These measurements are vital in understanding how different stimuli—like VR experiences—affect the human body. The data obtained can reflect a wide range of physiological processes, from brain activity to heart rate, and even emotional responses. When integrated with VR, these measurements can offer a deeper understanding of user engagement, stress levels, cognitive load, and emotional states.

These measurements are usually carried out using sensors that detect electrical, thermal, or mechanical signals generated by the body. The choice of measurement tools and techniques depends on the specific physiological process being monitored and the goals of the research.

10.4.2 Electroencephalography (EEG)

EEG is one of the most common methods for measuring brain activity. It involves placing electrodes on the scalp to detect electrical activity produced by the firing of neurons in the brain. EEG is particularly useful in VR research for analyzing how different virtual environments and tasks influence cognitive states such as attention, relaxation, and mental workload.

The EEG signals are characterized by their frequency bands, commonly referred to as delta, theta, alpha, beta, and gamma waves, each associated with different brain states. For example, alpha waves are often linked to relaxation, while beta waves are associated with active thinking and focus. By monitoring these signals during VR experiences, researchers can infer the user's cognitive and emotional state, allowing for the customization of VR content to enhance user experience.

EEG systems used in VR research can range from clinical-grade devices to more portable, consumer-friendly headsets. These devices are typically synchronized with the VR system to correlate brain activity with specific events in the virtual environment, providing real-time insights into how the brain responds to VR stimuli [26, 27].

10.4.3 Pulse Rate

Pulse rate, or heart rate, is another crucial physiological measurement, indicating the number of heartbeats per minute. It serves as a primary indicator of cardiovascular health and can also reflect a person's emotional and physiological state. In VR research, pulse rate monitoring is often used to assess stress, anxiety, excitement, or relaxation in response to different virtual scenarios.

Pulse rate can be measured using photoplethysmography (PPG), a non-invasive technique that uses light to detect blood volume changes in the microvascular bed of tissue. Devices such as smartwatches, fitness trackers, and specialized medical equipment can provide accurate pulse rate readings. In VR setups, these devices can be synchronized with the VR system to track how the user's heart rate changes in response to virtual events, offering insights into their emotional and physiological engagement with the experience.

10.4.4 Perspiration Rate

Perspiration rate, or skin conductance, is a measure of the skin's ability to conduct electricity, which varies with the amount of sweat produced. This physiological measure is directly linked to the sympathetic nervous system, which controls the body's fight-or-flight response. An increase in perspiration usually indicates heightened emotional arousal, whether due to stress, excitement, or fear.

In VR research, perspiration rate is typically measured using galvanic skin response (GSR) sensors. These sensors detect changes in electrical conductance on the skin surface, usually on the palms or fingertips, where sweat glands are most active. By monitoring perspiration during VR experiences, researchers can gauge the emotional intensity of the experience and determine which elements of the VR environment are most engaging or stressful.

10.4.5 Body Temperature

Body temperature is another physiological parameter that can reflect a person's state of arousal or stress. Typically, body temperature is stable, but it can fluctuate in response to environmental conditions, physical activity, and emotional stress. Monitoring body temperature in VR can provide insights into how immersive experiences affect the user's autonomic nervous system, which regulates body temperature among other functions.

Temperature sensors used in VR studies are often placed on the skin, where they can detect subtle changes in temperature that may indicate physiological responses to the virtual environment. For example, a decrease in peripheral skin temperature might suggest vasoconstriction due to stress, while an increase might indicate relaxation or comfort.

10.4.6 Facial Expression Analysis

Facial expressions are powerful indicators of a person's emotional state. They provide immediate, observable data on how an individual is feeling in response to stimuli. In VR research, facial expression analysis is used to understand how users emotionally engage with virtual content.

This analysis can be performed using cameras and software that track and interpret facial muscle movements. The software can identify expressions associated with emotions like happiness, surprise, fear, or disgust. By analyzing these expressions in real time, researchers can evaluate the emotional impact of VR experiences and make adjustments to improve user engagement and satisfaction.

10.4.7 Blinking

Blinking is a simple yet informative physiological response. The rate and pattern of blinking can indicate various cognitive and emotional states, such as concentration, fatigue, or stress. In VR research, blinking is often monitored to assess how different virtual environments affect user engagement and cognitive load.

Blink rate can be measured using eye-tracking technology, which is often integrated into VR headsets or sensor glasses. By tracking when and how often a user blinks, researchers can infer levels of attention and mental workload. For example, a decreased blink rate may indicate high concentration, while an increased rate might suggest fatigue or cognitive overload.

10.4.8 Conclusion

The integration of physiological measurements with VR technology offers a powerful toolset for understanding human responses to virtual environments. By monitoring EEG, pulse rate, perspiration rate, body temperature, facial expressions, and blinking, researchers can gain a comprehensive view of how users interact with and are affected by VR experiences. These measurements not only provide insights into the physiological and emotional states of users but also guide the development of more immersive and user-centered VR applications. As VR technology continues to evolve, the role of physiological measurements in enhancing and personalizing VR experiences will undoubtedly become even more critical, paving the way for innovations in fields ranging from healthcare to entertainment.

10.5 The Meaning of VR Experiments Combined with Physiological Measurements for E-Learning

In recent years, there has been a significant push towards integrating innovative technologies into educational environments to enhance learning experiences. One of the most promising areas of this development is the use of Virtual Reality (VR) in e-learning, particularly when combined with physiological measurements. This section explores the potential of VR in e-learning, emphasizing the role of Problem-Based Learning (PBL) and the importance of understanding the mental reactions that occur during such immersive educational experiences.

10.5.1 The Role of PBL in E-Learning

Problem-Based Learning (PBL) is an instructional method that focuses on engaging students in solving real-world problems as a way to acquire knowledge and develop critical thinking skills. Traditionally, PBL has been most effective in physical classroom settings, where students can actively collaborate, communicate, and interact with their peers and instructors. This interaction fosters an active learning environment, which is essential for deep learning and retention of knowledge.

However, with the rise of e-learning, there is a growing need to adapt PBL to virtual environments. If successfully implemented, PBL in e-learning could offer the same benefits of active learning while overcoming geographical barriers, geopolitical challenges, and health-related restrictions, such as those posed by pandemic situations. E-learning platforms that incorporate PBL could enable students from diverse locations to collaborate on problem-solving tasks in a virtual space, thereby broadening access to quality education.

10.5.2 The Potential of VR in E-Learning

Virtual Reality (VR) has the potential to revolutionize e-learning by providing immersive environments that mimic the dynamics of traditional classrooms. In a VR setting, students can interact with 3D objects, explore virtual environments, and engage in simulations that are otherwise impossible in a conventional online learning platform. For instance, VR can transport students into a virtual laboratory where they can conduct experiments, or into a historical setting where they can witness events firsthand.

The immersive nature of VR can make learning more engaging and effective. It enhances students' sense of presence, making them feel as though they are physically present in the virtual environment. This heightened sense of presence can lead to better retention of information, increased motivation, and greater engagement with the learning material.

10.5.3 Combining VR with Physiological Measurements

While VR can create an engaging e-learning environment, understanding the mental reactions of participants—both students and teachers—during VR experiences is crucial. This is where the combination of VR with physiological measurements becomes valuable. By monitoring physiological indicators such as EEG, pulse rate, perspiration rate, and facial expressions, educators and researchers can gain insights into the cognitive and emotional states of learners during VR-based e-learning sessions.

For example, EEG measurements can reveal how focused or relaxed a student is during a VR activity. Pulse rate and perspiration rate can indicate levels of stress or excitement, while facial expressions can provide real-time feedback on how students are emotionally responding to the learning content. Blinking patterns, as discussed earlier, can also offer clues about cognitive load and attention.

These physiological measurements can help educators tailor VR content to better suit the needs of learners. For instance, if physiological data indicates that a particular VR simulation is causing high levels of stress, the content can be adjusted to make it less overwhelming. Alternatively, if students are not sufficiently engaged, the VR environment can be modified to be more stimulating.

10.5.4 Establishing New Classroom Systems

The integration of VR and physiological measurements in e-learning could lead to the development of new, more effective classroom systems. These systems would not only facilitate active learning but also allow for real-time monitoring and adjustment of the learning experience based on students' physiological responses. This adaptive approach to education could result in more personalized learning experiences, where the content and pacing are tailored to the individual needs of each student.

Moreover, such systems could help overcome some of the limitations of traditional e-learning. For example, in a purely online setting, it can be difficult for instructors to gauge how well students are understanding the material or how engaged they are. By using physiological measurements, instructors can obtain objective data on student engagement and comprehension, enabling them to intervene when necessary to keep students on track.

In the context of PBL, this approach could enhance the collaborative aspect of learning by ensuring that all students are fully engaged and contributing to the problem-solving process. It could also help identify any cognitive or emotional barriers that may be hindering a student's ability to participate effectively, allowing for timely support and intervention.

10.5.5 Conclusion

The combination of VR and physiological measurements represents a significant advancement in the field of e-learning. By providing immersive, engaging environments and real-time insights into student engagement and cognitive states, this approach has the potential to transform PBL in e-learning. It offers a way to bring the benefits of active learning into the virtual space, making it possible to create dynamic, interactive learning experiences that transcend geographical and physical limitations. As we continue to explore and refine these technologies, we move closer to establishing new classroom systems that are not only effective but also adaptable to the diverse needs of learners worldwide.

10.6 Examples of Research Using Eye Potential Sensors—Insights from Our Studies

Eye potential sensors, particularly electrooculography (EOG) sensors, have emerged as powerful tools in physiological research, offering a window into the cognitive and emotional states of individuals based on their eye movements. In our ongoing research, we have extensively utilized these sensors to explore various aspects of human interaction with digital environments, particularly in the context of Virtual Reality (VR) and e-learning applications. This section provides an overview of some of our key studies that have employed eye potential sensors, highlighting their applications and the insights gained.

10.6.1 Study 1: Eye Movement Analysis in VR Environments

One of our foundational studies involved using EOG sensors to monitor eye movements in participants navigating complex VR environments. The primary objective was to understand how different VR settings influenced visual attention and cognitive load. Participants were equipped with VR headsets integrated with EOG sensors, which allowed us to track their eye movements in real-time as they interacted with various virtual scenarios.

The study revealed significant variations in eye movement patterns based on the complexity of the virtual environment. For instance, in highly detailed and dynamic scenes, participants exhibited more frequent and rapid eye movements, indicative of increased cognitive load as they attempted to process the multitude of visual stimuli. In contrast, simpler environments resulted in more stable and focused eye movements, suggesting lower cognitive demand. These findings underscore the importance of designing VR environments that are cognitively engaging yet not overwhelming, particularly in educational contexts where the balance between stimulation and comprehension is crucial.

10.6.2 Study 2: Blinking Patterns and Cognitive States

In another study, we investigated the relationship between blinking patterns and cognitive states using EOG sensors. Blinking is a subtle yet informative physiological response that can indicate levels of concentration, fatigue, and stress. By analyzing blinking frequency and duration, we aimed to assess how these factors varied in response to different tasks within a VR environment.

The results demonstrated that blinking rates decreased during tasks requiring high levels of concentration, such as problem-solving activities or navigating through complex virtual mazes. Conversely, tasks that were less cognitively demanding, such as passive observation of a virtual landscape, were associated with higher blinking rates, indicative of a more relaxed state. These findings suggest that blinking patterns can serve as a valuable metric for assessing cognitive load in real-time, providing insights into the mental demands placed on users during VR-based learning or training sessions.

10.6.3 Study 3: Emotional Responses in E-Learning Scenarios

Building on our previous research, we conducted a study to explore the emotional responses of students engaged in e-learning scenarios using VR, with a focus on the role of eye potential sensors in capturing these responses. The study involved students participating in interactive lessons delivered through VR headsets, with EOG sensors used to monitor their eye movements and blinking patterns.

The data revealed distinct differences in eye movement and blinking patterns corresponding to various emotional states. For example, lessons that included unexpected challenges or competitive elements triggered faster and more erratic eye movements, along with increased blinking, indicating heightened arousal and stress. In contrast, segments of the lesson designed to be calming or reassuring, such as guided meditation or soothing visuals, resulted in slower eye movements and reduced blinking, reflecting a state of relaxation.

These insights are particularly relevant for the design of e-learning platforms, as they highlight the potential for using eye potential sensors to tailor content delivery based on real-time emotional feedback. By adjusting the pacing, difficulty, or type of content in response to a learner's emotional state, educators can create more personalized and effective learning experiences.

10.6.4 Study 4: Eye Potential Sensors in Collaborative Virtual Environments

In a collaborative study, we examined the use of EOG sensors in multi-user VR environments, where participants were required to work together to solve complex problems. The focus was on understanding how eye movements and blinking patterns differed in collaborative versus individual settings, and what this could reveal about group dynamics and communication.

The study found that in collaborative settings, participants exhibited more synchronized eye movements, particularly during tasks that required close coordination or shared focus on specific virtual objects. Additionally, blinking patterns were more aligned among group members, suggesting a level of cognitive and emotional synchronization. This phenomenon, often referred to as "social mirroring," indicates that eye potential sensors could be valuable tools for studying group behavior and enhancing the design of collaborative VR platforms.

10.6.5 Study 5: Application of Eye Potential Sensors in Remote Education

The final study discussed in this section explores the application of eye potential sensors in remote education settings, where students participated in VR-based lessons from different geographical locations. The aim was to assess how eye movement data could be used to monitor and improve engagement in remote learning environments, which are becoming increasingly relevant in the context of global educational trends.

The findings revealed that eye potential sensors provided reliable data on student engagement, which was particularly useful for remote educators who lacked the ability to observe students directly. By analyzing the patterns of eye movements and blinking, educators could identify when students were losing focus or becoming disengaged, allowing them to intervene with targeted prompts or adjustments to the lesson content. This approach has significant implications for the future of remote education, offering a means to maintain high levels of student engagement even in virtual classrooms.

10.6.6 Conclusion

These examples from our research demonstrate the versatility and effectiveness of eye potential sensors in studying human interaction with digital environments, particularly in the context of VR and e-learning. By providing real-time data on eye movements and blinking patterns, these sensors offer valuable insights into cognitive load, emotional states, and group dynamics, which can be used to enhance the design and implementation of VR-based educational platforms. As we continue to explore the potential of these technologies, it is clear that eye potential sensors will play a crucial role in the development of personalized, adaptive learning experiences that respond to the needs of individual learners in real-time.

10.7 Final Conclusion

This chapter aims to contribute to the burgeoning field of VR research by providing an in-depth look at the methods and technologies employed to capture physiological data in virtual environments. Through the lens of a specific experiment and a review of related studies, we seek to illuminate the potential of combining VR with advanced sensing technologies for enhanced user experience research. Our discussion extends beyond the technical implementation to consider the implications of these findings for the design and development of VR applications, highlighting the pivotal role of physiological data in crafting immersive and responsive virtual worlds.

References

1. Mazuryk, T., Gervautz, M.: History, applications, technology and future. Virtual Reality. **72**(4), 486–497 (1996)
2. Kurland, E.: History of VR. In: Virtual Reality Filmmaking, pp. 7–17. Routledge (2017)
3. Gigante, M.A.: Virtual reality: definitions, history and applications. In: Virtual Reality Systems, pp. 3–14. Academic (1993)
4. Trinidad-Fernández, M., Bossavit, B., Salgado-Fernández, J., Abbate-Chica, S., Fernández-Leiva, A.J., Cuesta-Vargas, A.I.: Head-mounted display for clinical evaluation of neck movement validation with meta quest 2. Sensors. **23**(6), 3077 (2023)
5. Wang, C.: Developments of sensing technology and its applications in virtual reality. Highlights Sci. Eng. Technol. **39**, 95–102 (2023)
6. Raymer, E., MacDermott, Á., Akinbi, A.: Virtual reality forensics: forensic analysis of Meta Quest 2. Forensic Sci. Int. Digit. Investig. **47**, 301658 (2023)
7. Stein, C.: Virtual reality design: how upcoming head-mounted displays change design paradigms of virtual reality worlds. MediaTropes. **6**(1) (2016)
8. Keshner, E.A., Weiss, P.T., Geifman, D., Raban, D.: Tracking the evolution of virtual reality applications to rehabilitation as a field of study. J. Neuroeng. Rehab. **16**, 1–15 (2019)

9. Muñoz-Saavedra, L., Miró-Amarante, L., Domínguez-Morales, M.: Augmented and virtual reality evolution and future tendency. Appl. Sci. **10**(1), 322 (2020)
10. Noah, N., Das, S.: Exploring evolution of augmented and virtual reality education space in 2020 through systematic literature review. Comput. Anim. Virtual Worlds. **32**(3–4), e2020 (2021)
11. Cameirão, M.S., Bermúdez i Badia, S., Oller, E.D., Verschure, P.F.M.J.: The rehabilitation gaming system: a review. Adv. Technol. Rehab., 65–83 (2009)
12. Mantovani, F.: 12 VR learning: Potential and challenges for the use of 3D environments in education and training. Towards Cyberpsychol. Mind Cogn. Soc. Internet Age. **2**(207) (2001)
13. Hsieh, M.-C., Lee, J.-J.: Preliminary study of VR and AR applications in medical and healthcare education. J. Nurs. Health Stud. **3**(1), 1 (2018)
14. Smit, F.A., van Liere, R., Fröhlich, B.: An image-warping VR-architecture: design, implementation and applications. In: Proceedings of the 2008 ACM Symposium on Virtual Reality Software and Technology, pp. 115–122 (2008)
15. McVeigh-Schultz, J., Kolesnichenko, A., Isbister, K.: Shaping pro-social interaction in VR: an emerging design framework. In: Proceedings of the 2019 CHI Conference on Human Factors in Computing Systems, pp. 1–12 (2019)
16. Jung, T., Claudia Tom Dieck, M., Moorhouse, N., Tom Dieck, D.: Tourists' experience of virtual reality applications. In: 2017 IEEE International Conference on Consumer Electronics (ICCE), pp. 208–210. IEEE (2017)
17. Milosevic, I.: Fight-or-flight response. Phobias Psychol. Irrational Fear. **196**, 179 (2015)
18. Appelhans, B.M., Luecken, L.J.: Heart rate variability as an index of regulated emotional responding. Rev. Gen. Psychol. **10**(3), 229–240 (2006)
19. Wientjes, C.J.E.: Respiration in psychophysiology: methods and applications. Biol. Psychol. **34**(2–3), 179–203 (1992)
20. Khalfa, S., Isabelle, P., Jean-Pierre, B., Manon, R.: Event-related skin conductance responses to musical emotions in humans. Neurosci. Lett. **328**(2), 145–149 (2002)
21. Kanematsu, H., Barry, D.M., Shirai, T., Kawaguchi, M., Ogawa, N., Yajima, K., Nakahira, K.T., Kobayashi, T., Yoshitake, M.: Measurements of eye movement and teachers' concentration during the preparation of teaching materials. Proc. Comput. Sci. **159**, 1499–1506 (2019)
22. Kanematsu, H., Ogawa, N., Shirai, T., Kawaguchi, M., Kobayashi, T., Barry, D.M.: Blinking eyes behaviors and face temperatures of students in Youtube lessons—for the future e-learning class. Proc. Comput. Sci. **96**, 1619–1626 (2016)
23. Yajima, K., Kanematsu, H., Barry, D.M., Shirai, T., Kawaguchi, M., Ogawa, N., Nakahira, K.T., Suzuki, S.-n., Kobayashi, T., Yoshitake, M.: Application of biological information from eye blinking to mutual communication for e-learning: results of pbl activities for students. Proc. Comput. Sci. **176**, 3029–3036 (2020)
24. Dharmawansa, A.D., Fukumura, Y., Kanematsu, H., Kobayashi, T., Ogawa, N., Barry, D.M.: Introducing eye blink of a student to the virtual world and evaluating the affection of the eye blinking during the e-learning. Proc. Comput. Sci. **35**, 1229–1238 (2014)
25. Barry, D.M., Ogawa, N., Dharmawansa, A., Kanematsu, H., Fukumura, Y., Shirai, T., Yajima, K., Kobayashi, T.: Evaluation for students' learning manner using eye blinking system in metaverse. Proc. Comput. Sci. **60**, 1195–1204 (2015)
26. Kanematsu, H., Barry, D.M., Ogawa, N., Nakahira, K.T., Yoshitake, M., Shirai, T., Kawaguchi, M., Kobayashi, T., Yajima, K.: Some psychological responses measured by a commercial electrooculography sensor and its applicability. Proc. Comput. Sci. **126**, 1014–1022 (2018)
27. Kanematsu, H., Barry, D.M., Ogawa, N., Yajima, K., Nakahira, K.T., Suzuki, S.-n., Kato, T., Shirai, T., Kawaguchi, M., Yoshitake, M.: Evaluation methods and differences between three dimensional and two dimensional movies by physiological measurements using a commercial 6-axis sensor. Proc. Comput. Sci. **225**, 4631–4639 (2023)

Chapter 11
For the Future: Fusion of Real and Virtual Worlds

Hideyuki Kanematsu and Dana M. Barry

Abstract The integration of Digital Twin technology with Augmented Reality (AR) is rapidly transforming industries by bridging the gap between the physical and digital worlds. This chapter explores the evolution of virtual tools and their role in creating and managing Digital Twins, which serve as dynamic, real-time replicas of physical entities. The chapter delves into the distinctions between AR, VR, and Mixed Reality (MR), and their applications across various industries such as manufacturing, healthcare, and smart cities. Through case studies and practical examples, the chapter illustrates how AR enhances real-world experiences by overlaying digital information onto physical environments, leading to improved decision-making, efficiency, and innovation. Additionally, the chapter addresses the technical and ethical challenges of combining AR with Digital Twin technology, such as data integration, scalability, security, and privacy concerns. It also highlights the opportunities for innovation in sectors like automotive, aerospace, and healthcare, and discusses future trends that will shape the continued evolution of these technologies. The chapter concludes by emphasizing the transformative potential of AR and Digital Twin technology in revolutionizing industry practices and driving sustainable innovation.

Keywords Digital twin technology · Augmented reality (AR) · Virtual reality (VR) · Industry 4.0 · Real-time data integration

H. Kanematsu (✉)
National Institute of Technology, Suzuka College, Suzuka-shi, Mie, Japan
e-mail: hideyuki.kanematsu@bioenglab.org

D. M. Barry
Clarkson University, Potsdam, NY, USA
e-mail: dbarry@clarkson.edu

D. M. Barry, H. Kanematsu (eds.), *Applications of Metaverse and Virtual Reality to Creative Education and Industry*, Intelligent Systems Reference Library 267, https://doi.org/10.1007/978-981-96-3341-8_11

11.1 Introduction

Integrating real and virtual worlds has opened new avenues for technological innovation, fundamentally transforming how industries operate and interact with their environments. With the rise of Industry 4.0, the convergence of digital and physical realms has become increasingly significant, leading to the development of cutting-edge technologies like Augmented Reality (AR) and Digital Twin. AR enhances our perception of reality by overlaying digital information onto the physical world. At the same time, Digital Twin technology creates precise virtual models of real-world entities, enabling simulations and real-time data analysis [1–9].

This chapter explores the future scope of combining AR with Digital Twin technology, focusing on how these virtual tools are applied across various industries. By examining concrete examples and addressing both the challenges and opportunities presented by this convergence, we aim to shed light on the transformative potential of AR and Digital Twin technologies. These innovations are poised to revolutionize manufacturing and healthcare sectors, offering unprecedented possibilities for improving efficiency, safety, and decision-making processes.

In the following sections, we will delve into the specifics of AR and Digital Twin applications, providing a comprehensive overview of their current state and for the future.

11.2 Understanding Augmented Reality (AR) and Virtual Reality (VR)

Definitions and Distinctions Between AR, VR, and Mixed Reality (MR)
Augmented Reality (AR) and Virtual Reality (VR) are two emerging technologies rapidly transforming how we interact with digital content and the physical world. While AR and VR are immersive technologies, they differ significantly in their approach and application.

Virtual Reality (VR) creates a completely immersive digital environment, effectively transporting users into a virtual world entirely separate from their physical surroundings. VR experiences require specialized equipment such as head-mounted displays (HMDs), which completely obscure the user's view of the real world, replacing it with a computer-generated environment. This technology is often used in gaming, training simulations, and virtual tours, where users can interact with the virtual environment in a real way.

Augmented Reality (AR), on the other hand, overlays digital content onto the real world, enhancing the user's perception of their physical environment. Unlike VR, AR does not require a fully immersive setup; it can be experienced through devices like smartphones, tablets, or AR glasses. The technology superimposes images, videos, and other digital information onto the user's view of the real world, creating a blended experience. AR is commonly used in applications like navigation

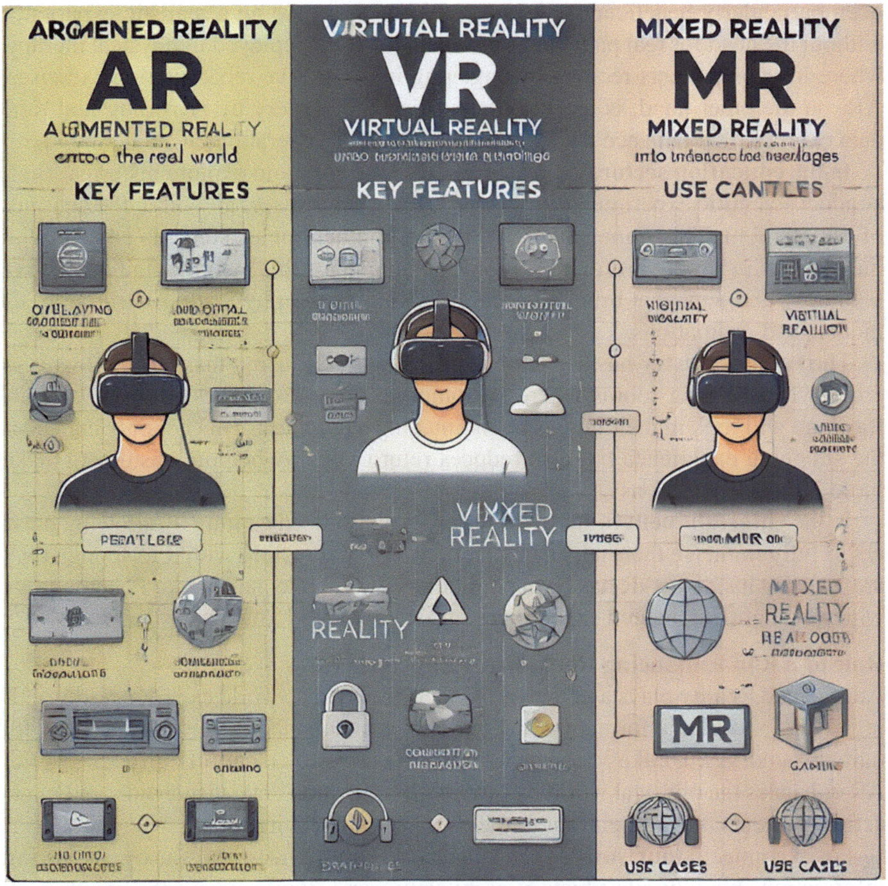

Fig. 11.1 The schematic diagram of the relation among AR, VR and MR

systems, interactive marketing, and education, where real-time information is provided without isolating the user from their surroundings.

Mixed Reality (MR) is often seen as a blend of AR and VR, where digital and physical worlds are merged in real-time. In MR, users can interact with digital objects as if they exist in their physical space, with these objects responding to the user's actions and the environment. This creates a more seamless integration between virtual and real-world experiences, allowing for complex simulations and interactions that are not possible with AR or VR alone [10] (Fig. 11.1).

Current Applications of AR and VR in Various Industries

Both AR and VR technologies have found extensive applications across a range of industries, transforming traditional methods and creating new opportunities for innovation.

In the **healthcare industry,** VR is used for surgical simulations, allowing surgeons to practice complex procedures in a risk-free virtual environment. This has been particularly beneficial for training purposes, providing a realistic experience without the need for real patients. Additionally, VR is employed in physical therapy, where it helps patients recover by engaging in immersive rehabilitation exercises. AR, on the other hand, is used in diagnostics and surgery by providing real-time data overlays that enhance the surgeon's view during operations.

In the **education sector,** VR offers students the ability to explore historical sites, conduct scientific experiments, or even travel through space, all within the confines of a classroom. This immersive learning environment helps students engage with the material in a more profound way. AR complements this by providing interactive textbooks and apps that bring subjects like biology and physics to life through 3D models and simulations.

The **retail industry** has also embraced AR and VR, using these technologies to create virtual fitting rooms where customers can try on clothes or visualize how furniture will look in their homes before making a purchase. This not only enhances the shopping experience but also reduces return rates by helping customers make more informed decisions.

In the **entertainment industry,** VR has revolutionized gaming by allowing players to fully immerse themselves in a game world, interacting with it as though they are actually inside it. AR has been popularized by mobile games like Pokémon Go, which brought augmented experiences into the real world on a massive scale.

Role of AR in Enhancing Real-World Experiences

Augmented Reality plays a unique role in enhancing real-world experiences by adding layers of digital information that enrich our understanding and interaction with our surroundings. Unlike VR, which transports users to an entirely different world, AR enhances the physical world by providing contextual data, visual aids, and interactive elements that are relevant to the user's current environment.

For example, in **tourism,** AR can enhance a visit to a historical site by overlaying information about the site's history, architecture, and significant events directly onto the visitor's view. This turns a passive visit into an interactive learning experience. Similarly, in **navigation,** AR apps can overlay directions, points of interest, and other navigational aids onto the real world, making it easier for users to find their way in unfamiliar locations [11].

In **maintenance and repair,** AR can provide technicians with real-time visual instructions and data overlays that guide them through complex procedures. This reduces the likelihood of errors and allows even less experienced workers to perform tasks with greater confidence [12].

In the **art and culture sector,** AR is used to create interactive exhibits in museums and galleries, where visitors can see how ancient artifacts looked in their original context or explore detailed 3D reconstructions of historical buildings [13, 14].

Overall, AR enhances real-world experiences by making information more accessible and interactive, thereby bridging the gap between the digital and physical worlds in a way that is seamless and intuitive.

In conclusion, Augmented Reality (AR) and Virtual Reality (VR) are transformative technologies with distinct characteristics and applications. While VR offers a fully immersive experience by transporting users to a completely virtual environment, AR enhances the real world by overlaying digital information onto physical surroundings. Both technologies are being utilized across various industries, from healthcare and education to retail and entertainment, offering new ways to interact with digital content and the physical world. As these technologies continue to evolve, their potential to enhance our lives and work environments will only grow, making them integral to future innovations.

11.3 Digital Twin: The Convergence of Physical and Digital Worlds

What Is a Digital Twin?

Digital Twin is a cutting-edge technology that represents the convergence of physical and digital worlds by creating a virtual model or replica of a physical object, process, or system. This digital replica enables real-time monitoring, simulation, and analysis, offering a comprehensive view of the physical entity it mirrors. Essentially, a Digital Twin serves as a bridge between the physical and digital realms, allowing for the seamless flow of information between the two.

The concept of the Digital Twin emerged as a key innovation in Industry 4.0, where the integration of advanced technologies like IoT (Internet of Things), AI (Artificial Intelligence), and big data analytics became crucial. By utilizing sensors and data collection methods, the Digital Twin continuously receives updates from its physical counterpart, ensuring that it remains an accurate representation. This dynamic interaction allows businesses and industries to optimize performance, predict potential issues, and make data-driven decisions with unprecedented precision.

Concrete Examples of Digital Twin Technology

Example 1: Manufacturing Industry In the manufacturing sector, Digital Twin technology is revolutionizing the way products are designed, produced, and maintained. For instance, in a car manufacturing plant, every machine, assembly line, and even the end product can have its own Digital Twin. These digital replicas allow engineers to simulate production processes, identify bottlenecks, and test modifications in a virtual environment before implementing them on the factory floor. By doing so, manufacturers can reduce downtime, improve efficiency, and ensure quality control [15, 16].

Example 2: Smart Cities Smart cities utilize Digital Twin technology to create virtual models of entire urban environments. These models integrate data from various sources, such as traffic systems, energy grids, and public services, to optimize city management. For example, a Digital Twin of a city's transportation network can simulate traffic flow, predict congestion, and suggest real-time adjustments to traffic signals. This helps in reducing traffic jams, lowering emissions, and improving overall urban mobility [17, 18].

Example 3: Healthcare and Personalized Medicine In healthcare, Digital Twins are being used to create personalized models of patients. These models can simulate how a patient's body might respond to different treatments or surgeries. For instance, before performing a complex surgery, a surgeon can use a patient's Digital Twin to practice the procedure in a virtual environment, assessing potential risks and outcomes. This technology is also being used to monitor chronic diseases, where the Digital Twin of a patient continuously updates with real-time health data, allowing doctors to make informed decisions and provide personalized care [19–23].

The Meaning and Significance of Digital Twin Technology

The significance of Digital Twin technology lies in its ability to transform how industries operate by providing a deeper understanding of physical systems through their digital counterparts. One of the most important aspects of this technology is its predictive capability. By analyzing real-time data and historical trends, Digital Twins can predict potential failures, maintenance needs, or performance issues before they occur. This predictive maintenance can save companies significant amounts of money by preventing unplanned downtime and extending the lifespan of machinery.

Another critical aspect is the role of Digital Twins in innovation. By simulating different scenarios in a virtual environment, companies can experiment with new ideas, processes, or designs without the risk and cost associated with physical testing. This accelerates the innovation cycle, allowing businesses to bring new products to market faster and with greater confidence in their reliability.

Moreover, Digital Twins facilitate better decision-making by providing a comprehensive view of the entire system or process. Whether it's a single machine or a complex urban infrastructure, having a digital replica that continuously updates in real-time enables decision-makers to see the full picture and make informed choices based on accurate, up-to-date information.

Concrete Application: Uniting the Virtual and Real Worlds

The true potential of Digital Twin technology is realized when it is used to unite the virtual world with the real world, creating a seamless integration that enhances both environments.

Application 1: Integrated Building Management In modern building management, Digital Twins are used to create virtual models of entire buildings, integrating data from HVAC systems, lighting, security, and occupancy sensors. These Digital Twins allow building managers to monitor and control the building's environment in real-time, optimizing energy usage, ensuring safety, and improving occupant comfort. For instance, if a room is unoccupied, the Digital Twin can automatically adjust the temperature and lighting to save energy, and if a security breach is detected, it can simulate potential responses and suggest the most effective course of action.

Application 2: Autonomous Vehicles In the automotive industry, Digital Twins play a crucial role in the development and operation of autonomous vehicles.

Each autonomous vehicle has a Digital Twin that collects data from sensors on the vehicle and its surroundings. This digital replica allows for continuous monitoring and updates, helping the vehicle to navigate safely and efficiently. The Digital Twin can simulate different driving scenarios, allowing the vehicle's AI to learn and improve its decision-making capabilities. This integration of the virtual and real worlds ensures that autonomous vehicles can operate safely in complex environments.

Application 3: Virtual Power Plants In the energy sector, Digital Twins are being used to create virtual power plants (VPPs), which are aggregations of decentralized energy resources such as solar panels, wind turbines, and battery storage systems. The Digital Twin of a VPP integrates data from all these sources to optimize the generation, storage, and distribution of energy. It can predict energy demand, adjust the output from various sources in real-time, and ensure the stability of the grid. This not only enhances the efficiency of energy production but also supports the integration of renewable energy sources into the grid, contributing to a more sustainable energy future.

Digital Twin technology represents a significant step forward in the convergence of physical and digital worlds. By creating dynamic, real-time digital replicas of physical entities, Digital Twins enable industries to optimize operations, predict and prevent issues, and innovate more effectively. The applications of this technology are vast and varied, from manufacturing and healthcare to smart cities and energy management. As Digital Twin technology continues to evolve, its ability to unite the virtual and real worlds will unlock new possibilities for efficiency, sustainability, and innovation across multiple sectors.

11.4 Virtual Tools Applied to Digital Twin Technology

11.4.1 Overview of Virtual Tools Used in Creating and Managing Digital Twins

Digital Twin technology relies heavily on virtual tools to create, manage, and analyze digital replicas of physical systems. These tools provide the computational power and flexibility needed to simulate real-world scenarios, collect and process data, and visualize complex processes. Virtual tools are integral to the design, development, and continuous updating of Digital Twins, enabling them to function as accurate and dynamic models of their physical counterparts.

The primary role of virtual tools in Digital Twin technology is to bridge the gap between the physical and digital worlds. They allow for the seamless integration of real-time data from sensors and IoT devices with advanced computational models. These tools support the entire lifecycle of a Digital Twin, from initial creation to ongoing management, ensuring that the digital replica remains a true reflection of the physical system it represents.

11.4.2 Examples of Virtual Tools and Platforms: Unity, MATLAB, and Others

Unity Unity is a versatile platform that is widely used in the creation of Digital Twins due to its powerful 3D rendering capabilities and real-time simulation features. Originally developed for game design, Unity has found extensive applications in various industries, including automotive, manufacturing, and construction, where it is used to create highly detailed and interactive Digital Twins. Unity's ability to integrate with IoT devices and other data sources makes it an ideal tool for visualizing and interacting with complex systems in a virtual environment [24–28].

MATLAB MATLAB is a high-level programming environment that is particularly useful for mathematical modeling, simulation, and data analysis, all of which are crucial components of Digital Twin technology. MATLAB's Simulink platform allows for the simulation of dynamic systems, making it an excellent tool for designing and testing Digital Twins in fields such as aerospace, robotics, and energy. The platform's extensive library of toolboxes and its ability to handle large datasets make it indispensable for engineers and scientists working with Digital Twins.

Other Tools Several other virtual tools and platforms play a significant role in the development and management of Digital Twins:

- **ANSYS:** Known for its robust engineering simulation capabilities, ANSYS is frequently used in the design and optimization of Digital Twins, especially in industries like aerospace and automotive. Its ability to simulate complex physical phenomena, such as fluid dynamics and structural mechanics, makes it a critical tool for ensuring that Digital Twins accurately reflect their physical counterparts.
- **ThingWorx:** ThingWorx is an IoT platform specifically designed to support the creation and management of Digital Twins. It provides tools for data collection, analytics, and visualization, allowing for real-time monitoring and optimization of physical assets. ThingWorx is particularly popular in manufacturing and industrial IoT applications.
- **Autodesk Revit:** In the construction and architecture industries, Autodesk Revit is a leading tool for creating Digital Twins of buildings and infrastructure. It allows for detailed 3D modeling and integrates with BIM (Building Information Modeling) workflows, enabling architects and engineers to visualize and manage every aspect of a building's lifecycle.

11.4.3 Case Studies: Successful Implementation of Digital Twins Using Virtual Tools

Case Study 1: Unity in the Automotive Industry One of the most notable examples of Unity's application in Digital Twin technology is its use in the automotive industry. A leading car manufacturer employed Unity to create a Digital Twin of its entire production line. By simulating the production process in a virtual environment, the company was able to identify inefficiencies and optimize workflow before implementing changes in the physical plant. This proactive approach saved millions in potential downtime and retooling costs. Unity's real-time visualization capabilities also allowed engineers to test different scenarios, such as the introduction of new machinery or process changes, in a risk-free environment.

Case Study 2: MATLAB in Aerospace Engineering MATLAB, combined with Simulink, has been instrumental in the aerospace industry, particularly in the development of Digital Twins for aircraft engines. A prominent aerospace company used MATLAB to create a Digital Twin of its jet engine, integrating real-time data from sensors with simulation models to monitor engine performance. This Digital Twin allowed engineers to predict maintenance needs, optimize fuel efficiency, and prevent potential failures. The use of MATLAB enabled the company to reduce unexpected engine downtime, thereby increasing the reliability and safety of its aircraft.

Case Study 3: ThingWorx in Industrial IoT ThingWorx has been successfully implemented in the manufacturing sector to create Digital Twins of industrial machinery. A major industrial equipment manufacturer used ThingWorx to develop a Digital Twin of its fleet of connected machines. The platform's real-time data analytics and visualization tools provided insights into machine performance, enabling predictive maintenance and reducing unplanned outages. As a result, the company was able to improve operational efficiency and extend the lifespan of its machinery, demonstrating the significant value that Digital Twin technology can bring to industrial IoT applications.

Virtual tools are essential in the creation and management of Digital Twins, providing the necessary computational power, simulation capabilities, and real-time data integration to bridge the physical and digital worlds. Platforms like Unity, MATLAB, ANSYS, ThingWorx, and Autodesk Revit play critical roles in enabling the development of accurate and functional Digital Twins across various industries. The successful implementation of Digital Twins using these tools, as demonstrated in the automotive, aerospace, and industrial IoT sectors, highlights the transformative potential of this technology. As Digital Twin technology continues to evolve, the integration of advanced virtual tools will remain at the forefront, driving innovation and enhancing the capabilities of digital and physical systems alike.

11.5 Concrete Examples of AR in Digital Twin Applications

11.5.1 Example 1: AR in Manufacturing—Visualizing Production Lines in Real-Time

In the manufacturing industry, Augmented Reality (AR) is revolutionizing the way production lines are monitored and managed by integrating with Digital Twin technology. A Digital Twin of a production line creates a virtual model that mirrors the physical environment in real-time. When combined with AR, this allows workers and managers to visualize and interact with the production process directly from their mobile devices or AR glasses (Fig. 11.2).

For example, an operator on the factory floor can use AR glasses to see a real-time overlay of the entire production line, including data on machine performance,

Fig. 11.2 The image for a concrete application of AR in Digital Twin

workflow efficiency, and potential bottlenecks. This visual representation enables quick identification of issues such as machinery malfunctions or delays in the production process. By seeing these problems in real-time and in context, workers can take immediate action to resolve them, reducing downtime and increasing overall efficiency.

Moreover, AR can be used to provide step-by-step guidance for maintenance or assembly tasks, which is particularly beneficial in complex manufacturing environments. Workers can see instructions and diagrams overlaid on the physical components they are working on, ensuring that tasks are completed accurately and efficiently. This integration of AR with Digital Twins not only enhances productivity but also improves safety by reducing the likelihood of errors [29, 30].

11.5.2 Example 2: AR in Healthcare—Real-Time Monitoring of Patient Data Through Digital Twins

In healthcare, AR combined with Digital Twin technology offers groundbreaking possibilities for patient care and treatment. A Digital Twin of a patient is a dynamic virtual model that continuously updates with real-time data from various sources, such as wearable devices, medical records, and diagnostic tests. This model provides a comprehensive and up-to-date representation of the patient's health.

When integrated with AR, healthcare professionals can visualize a patient's Digital Twin directly on their devices or through AR glasses during consultations or surgeries. For instance, a surgeon can use AR to project a patient's vital signs, imaging data, and other relevant information onto their field of view while performing an operation. This allows the surgeon to have all the necessary data at their fingertips without having to look away from the patient or the surgical site.

In another scenario, a doctor could use AR to examine a patient remotely. By accessing the patient's Digital Twin, the doctor can visualize real-time data such as heart rate, blood pressure, and oxygen levels. This is particularly useful in telemedicine, where physical interaction is limited, but a detailed understanding of the patient's condition is still required. AR enhances the ability to monitor, diagnose, and treat patients effectively, even from a distance [31, 32].

11.5.3 Example 3: AR in Smart Cities—Integrating Real-Time Urban Data into Digital Twin Models

Smart cities are increasingly relying on Digital Twin technology to manage urban environments more efficiently. A Digital Twin of a city integrates data from various sources, such as traffic systems, public utilities, and environmental sensors, to create a comprehensive virtual model of the urban landscape. When combined with

AR, this technology allows city planners, managers, and even residents to interact with the city's infrastructure in real-time.

For example, AR can be used to visualize traffic flows, public transportation schedules, and road conditions in real-time, directly on a user's smartphone or AR glasses. A city planner might use this technology to see how traffic is moving through the city and identify areas where congestion is likely to occur. They can then simulate different scenarios, such as adjusting traffic signals or rerouting traffic, to alleviate congestion before it becomes a problem.

In addition, AR can provide city residents with real-time information about their surroundings. For instance, an AR app could overlay data about air quality, noise levels, or public transportation options onto the user's view of the city. This empowers residents to make informed decisions about their daily activities, such as choosing a less polluted route for their commute or avoiding noisy areas.

Furthermore, AR can be used in urban planning and development projects. By integrating real-time data into a city's Digital Twin, planners can visualize proposed changes to the urban landscape, such as new buildings or infrastructure projects, in the context of the existing environment. This helps to assess the potential impact of these projects on traffic, utilities, and the overall livability of the city, leading to more informed and sustainable urban development.

The integration of Augmented Reality (AR) with Digital Twin technology is transforming industries by providing powerful tools for real-time visualization and interaction. In manufacturing, AR enhances the ability to monitor and manage production lines with greater efficiency and accuracy. In healthcare, it allows for more precise and informed patient care through real-time monitoring and data visualization. In smart cities, AR and Digital Twins work together to create more responsive and efficient urban environments, improving the quality of life for residents and enabling better city management. As AR technology continues to evolve, its applications in conjunction with Digital Twins will only expand, offering new possibilities for innovation and improvement across various sectors.

11.6 Challenges and Opportunities

11.6.1 Technical and Ethical Challenges in Combining AR with Digital Twin Technology

As promising as the combination of Augmented Reality (AR) and Digital Twin technology is, it is not without its challenges. These challenges span both technical and ethical domains, presenting significant hurdles that must be addressed to fully realize the potential of these technologies.

11.6.2 Technical Challenges

1. **Data Integration and Management:**
   - One of the primary technical challenges in combining AR with Digital Twin technology lies in the integration and management of vast amounts of data. Digital Twins rely on real-time data from various sources, including IoT devices, sensors, and databases, to accurately replicate physical systems in a virtual environment. Integrating this data with AR applications requires robust data processing capabilities and seamless interoperability between different systems. Ensuring that data is accurately synchronized and presented in real-time through AR interfaces is critical, but it can be technically complex and resource-intensive.
2. **Scalability and Performance:**
   - As Digital Twins and AR applications become more sophisticated, the demand for computational power and storage increases significantly. Scaling these technologies to handle complex systems, such as entire manufacturing plants or urban infrastructures, poses a considerable challenge. The performance of AR applications can be affected by the sheer volume of data being processed and rendered in real-time, leading to potential latency issues and reduced user experience quality. Ensuring that AR applications remain responsive and accurate, even as the underlying Digital Twin models grow in complexity, requires ongoing advancements in computing hardware and software optimization.
3. **Security and Privacy:**
   - The integration of AR with Digital Twins introduces new security and privacy concerns. Digital Twins often contain sensitive information about physical assets, processes, and environments, making them attractive targets for cyberattacks. Ensuring the security of this data, especially when it is transmitted and accessed through AR devices, is paramount. Moreover, AR applications often require access to real-time data from users' environments, raising concerns about privacy and data protection. Developing secure and privacy-conscious frameworks for AR and Digital Twin integration is crucial to prevent unauthorized access and misuse of sensitive information.
4. **Interoperability and Standards:**
   - Another significant challenge is the lack of standardized protocols and interoperability between different AR platforms and Digital Twin systems. The diversity of hardware, software, and data formats used in these technologies can lead to compatibility issues, making it difficult to create seamless, cross-platform experiences. Establishing industry-wide standards for AR and Digital Twin integration would facilitate greater interoperability, allowing

different systems to work together more effectively and enabling wider adoption across various sectors.

11.6.3 Ethical Challenges

1. **Data Ownership and Consent:**

    - The use of AR in conjunction with Digital Twins raises important ethical questions about data ownership and consent. As these technologies become more pervasive, individuals and organizations must navigate the complexities of who owns the data generated by Digital Twins and AR applications. For instance, if a Digital Twin includes data collected from multiple stakeholders, determining who has the right to access, modify, or share this data can be challenging. Ensuring that users provide informed consent for the use of their data in AR applications is also critical to maintaining ethical standards [33, 34].

2. **Bias and Representation:**

    - The algorithms and data used to create Digital Twins and drive AR applications may inadvertently introduce bias, leading to inaccurate or skewed representations of reality. For example, if the data used to train a Digital Twin model is incomplete or biased, the resulting AR experience may reinforce existing disparities or exclude certain perspectives. It is essential to recognize and address these biases to ensure that AR and Digital Twin technologies are inclusive and representative of diverse populations and contexts.

3. **Impact on Employment and Skills:**

    - The adoption of AR and Digital Twin technologies in industries such as manufacturing, healthcare, and urban planning has the potential to significantly impact employment and workforce dynamics. While these technologies can enhance efficiency and innovation, they may also lead to job displacement, particularly for roles that become automated or obsolete. Additionally, the growing reliance on AR and Digital Twins may create new skills gaps, requiring workers to acquire specialized knowledge and training. Addressing these ethical challenges involves balancing the benefits of technological advancement with the need to support workers and communities affected by these changes.

11.6.4 Opportunities for Innovation in Industries Such as Automotive, Aerospace, and Healthcare

Despite the challenges, the integration of AR with Digital Twin technology presents numerous opportunities for innovation across various industries. By leveraging the strengths of both technologies, organizations can unlock new possibilities for enhancing productivity, improving safety, and driving innovation.

Automotive Industry

1. **Enhanced Design and Prototyping:**
   - In the automotive industry, the combination of AR and Digital Twins offers transformative potential in the design and prototyping stages of vehicle development. Engineers can use AR to visualize and interact with Digital Twins of vehicles in real-time, enabling them to make rapid design iterations and test different configurations without the need for physical prototypes. This accelerates the development process, reduces costs, and allows for more innovative designs to be explored.
2. **Predictive Maintenance and Diagnostics:**
   - Digital Twins of vehicles, when integrated with AR, can provide real-time insights into the condition and performance of individual components. For example, a mechanic could use AR glasses to visualize the Digital Twin of a car's engine, overlaying diagnostic data and predictive maintenance alerts directly onto the physical engine. This enables more accurate and timely maintenance, reducing the likelihood of breakdowns and extending the lifespan of the vehicle.

Aerospace Industry

1. **Simulation and Training:**
   - In the aerospace industry, AR and Digital Twins are revolutionizing simulation and training programs. Pilots and technicians can interact with Digital Twins of aircraft through AR interfaces, allowing them to practice complex procedures, such as emergency protocols or maintenance tasks, in a safe and controlled environment. These simulations can replicate real-world conditions with high fidelity, providing valuable hands-on experience without the risks associated with actual flights or operations.
2. **Lifecycle Management:**
   - The aerospace industry can also benefit from the integration of AR and Digital Twins in lifecycle management. Digital Twins of aircraft can be continuously updated with real-time data from sensors, allowing for ongoing monitoring of the aircraft's condition throughout its service life. AR applications can provide maintenance crews with real-time guidance on repair procedures, ensur-

ing that the aircraft remains in optimal condition and reducing the time required for inspections and repairs.

Healthcare Industry

1. **Personalized Medicine:**

   • In healthcare, the integration of AR with Digital Twins holds promise for advancing personalized medicine. By creating Digital Twins of individual patients, healthcare providers can simulate different treatment scenarios and predict how a patient might respond to specific therapies. AR can then be used to visualize these simulations in real-time during consultations, allowing doctors to tailor treatments to the unique needs of each patient and improve outcomes.

2. **Surgical Planning and Assistance:**

   • AR and Digital Twin technology are also enhancing surgical planning and assistance. Surgeons can use AR to visualize a patient's Digital Twin during preoperative planning, allowing them to explore different surgical approaches and anticipate potential challenges. During surgery, AR can provide real-time guidance, overlaying critical information, such as anatomical structures and instrument trajectories, directly onto the surgical field. This improves precision and reduces the risk of complications.

11.6.5 Future Trends and Potential Developments in This Field

As AR and Digital Twin technologies continue to evolve, several key trends and developments are likely to shape their future applications and impact.

1. **Increased Adoption Across Industries:**

   • The adoption of AR and Digital Twin technologies is expected to accelerate across a wide range of industries. As the technical challenges are addressed and the benefits of these technologies become more widely recognized, more organizations will integrate AR and Digital Twins into their operations. This will lead to greater innovation, improved efficiency, and enhanced decision-making across sectors such as manufacturing, healthcare, transportation, and urban planning.

2. **Enhanced Real-Time Collaboration:**

   • The future of AR and Digital Twin technology will likely see a focus on enhancing real-time collaboration. As remote work and global collaboration become more prevalent, AR will enable teams to interact with Digital Twins from different locations, sharing insights and making decisions in real-time.

This will facilitate more effective collaboration, particularly in complex projects that require input from multiple stakeholders, such as international manufacturing operations or large-scale infrastructure projects.

3. **Advances in AI and Machine Learning:**
   - The integration of Artificial Intelligence (AI) and Machine Learning (ML) with AR and Digital Twin technologies will drive significant advancements in their capabilities. AI-powered Digital Twins will be able to analyze vast amounts of data, identify patterns, and make predictive recommendations in real-time. This will enhance the accuracy and effectiveness of AR applications, enabling more sophisticated simulations, diagnostics, and decision-making processes.
4. **Greater Focus on Sustainability:**
   - Sustainability is becoming an increasingly important consideration in the development and deployment of new technologies. AR and Digital Twins have the potential to contribute to more sustainable practices by optimizing resource use, reducing waste, and improving efficiency. For example, Digital Twins can be used to simulate the environmental impact of different manufacturing processes or urban development projects, helping organizations make more sustainable choices. AR can support these efforts by providing real-time insights into resource usage and environmental performance.
5. **Evolution of Standards and Interoperability:**
   - As the adoption of AR and Digital Twin technologies grows, there will be a greater emphasis on developing industry standards and ensuring interoperability between different systems. This will enable more seamless integration of AR and Digital Twins across different platforms and applications, making it easier for organizations to implement these technologies and maximize their benefits.

The combination of AR and Digital Twin technology presents both significant challenges and exciting opportunities. While technical and ethical challenges must be addressed, the potential for innovation across industries such as automotive, aerospace, and healthcare is immense. As these technologies continue to evolve, they will play an increasingly important role in shaping the future of various sectors, driving advancements in efficiency, sustainability,

11.7 Conclusions

The integration of Digital Twin technology with Augmented Reality (AR) has emerged as a pivotal force driving the convergence of the physical and digital worlds. Through this integration, industries across the spectrum—from manufacturing and aerospace to healthcare and urban planning—are experiencing transformative changes that enhance efficiency, accuracy, and innovation.

Digital Twins offer a comprehensive, real-time representation of physical systems, allowing for predictive maintenance, optimization of processes, and better decision-making. When paired with AR, these digital replicas become even more powerful, providing users with an intuitive, visual interface to interact with and manipulate complex data. This combination enables real-time visualization and simulation, which not only enhances the user experience but also opens up new possibilities for innovation and problem-solving.

Despite the numerous advantages, the fusion of AR and Digital Twin technologies is not without its challenges. Technical hurdles such as data integration, scalability, and security need to be addressed to ensure the effective implementation of these technologies. Additionally, ethical considerations, including data ownership, consent, and the potential impact on employment, must be carefully managed to ensure that the benefits of these innovations are realized equitably and responsibly.

Looking forward, the continued evolution of AR and Digital Twin technologies will likely be shaped by advances in AI, machine learning, and sustainability efforts. These technologies will become more widespread as they offer new opportunities for real-time collaboration, more sophisticated simulations, and more sustainable practices. The development of industry standards and greater interoperability will further facilitate the adoption of these technologies across various sectors.

In conclusion, the combination of AR and Digital Twin technology represents a significant leap forward in how industries operate and innovate. As these technologies continue to evolve, they will undoubtedly play a crucial role in shaping the future, driving advancements in efficiency, sustainability, and the seamless integration of the digital and physical realms.

References

1. Chen, K., Nadirsha, T.N.M., Lilith, N., Alam, S., Svensson, Å.: Tangible digital twin with shared visualization for collaborative air traffic management operations. Transport. Res. C Emerg. Technol. **161**, 104546 (2024)
2. Grübel, J., Thrash, T., Aguilar, L., Gath-Morad, M., Chatain, J., Sumner, R.W., Hölscher, C., Schinazi, V.R.: The hitchhiker's guide to fused twins: a review of access to digital twins in situ in smart cities. Remote Sens. **14**(13), 3095 (2022)
3. Guo, Y., Liu, L., Huang, W., Shen, M., Yi, X., Zhang, J., Shizhu, L.: Extending X-reality technologies to digital twin in cultural heritage risk management: a comparative evaluation from the perspective of situation awareness. Herit. Sci. **12**(1), 245 (2024)
4. Alizadehsalehi, S.: BIM/Digital Twin-Based Construction Progress Monitoring through Reality Capture to Extended Reality (DRX) (2020)
5. Chung, M., Kim, K.A., Kang, M.S.: The status of metaverse and digital twin technology development. Korean J. Artif. Intel. **10**(2), 19–24 (2022)
6. Xu, C., Zhang, L.: Application of XR-based virtuality-reality coexisting course. Intel. Automation Soft Comput. **31**(3) (2022)
7. Grübel, J.: Experiments as DTs. In: Handbook of Digital Twins, pp. 563–583. CRC Press (2024)

8. Çirkin, E.: Industry 4.0 and Sustainability Implications. PhD dissertation, Dokuz Eylul Universitesi (Turkey) (2021)
9. Pronk, L.J.: The Design of a Roadmap for the Integration of Remote Support into a Production Environment. Master's thesis, University of Twente (2023)
10. Kanematsu, H., Barry, D.M., Shirai, T., Ogawa, N., Yajima, K., Nakahira, K.T., Kawaguchi, M., Suzuki, S.-n., Kato, T., Yoshitake, M.: Three dimensional experiences using a commercially-integrated stand-alone head-mounted display and applying it to education. Proc. Comput. Sci. **207**, 4288–4295 (2022)
11. Wei, W.: Research progress on virtual reality (VR) and augmented reality (AR) in tourism and hospitality: a critical review of publications from 2000 to 2018. J. Hosp. Tour. Technol. **10**(4), 539–570 (2019)
12. Palmarini, R., Erkoyuncu, J.A., Roy, R., Torabmostaedi, H.: A systematic review of augmented reality applications in maintenance. Robot. Comput. Integr. Manuf. **49**, 215–228 (2018)
13. Loumos, G.E.O.R.G.E., Kargas, A.N.T.O.N.I.O.S., Varoutas, D.: Augmented and virtual reality technologies in cultural sector: exploring their usefulness and the perceived ease of use. JMC. **4**, 307–322 (2018)
14. Guazzaroni, G., Pillai, A.S. (eds.): Virtual and augmented reality in Education, Art, and Museums. IGI Global (2019)
15. Leng, J., Wang, D., Shen, W., Li, X., Liu, Q., Chen, X.: Digital twins-based smart manufacturing system design in Industry 4.0: a review. J. Manuf. Syst. **60**, 119–137 (2021)
16. Tao, F., Zhang, M., Nee, A.Y.C.: Digital Twin Driven Smart Manufacturing. Academic (2019)
17. Mohammadi, N., Taylor, J.E.: Smart city digital twins. In: 2017 IEEE Symposium Series on Computational Intelligence (SSCI), pp. 1–5. IEEE (2017)
18. Dembski, F., Wössner, U., Letzgus, M., Ruddat, M., Yamu, C.: Urban digital twins for smart cities and citizens: the case study of Herrenberg, Germany. Sustainability. **12**(6), 2307 (2020)
19. Erol, T., Mendi, A.F., Doğan, D.: The digital twin revolution in healthcare. In: 2020 4th International Symposium on Multidisciplinary Studies and Innovative Technologies (ISMSIT), pp. 1–7. IEEE (2020)
20. Cellina, M., Cè, M., Alì, M., Irmici, G., Ibba, S., Caloro, E., Fazzini, D., Oliva, G., Papa, S.: Digital twins: the new frontier for personalized medicine? Appl. Sci. **13**(13), 7940 (2023)
21. Boulos, K., Maged, N., Zhang, P.: Digital twins: from personalised medicine to precision public health. J. Pers. Med. **11**(8), 745 (2021)
22. Rivera, L.F., Jiménez, M., Angara, P., Villegas, N.M., Tamura, G., Müller, H.A.: Towards continuous monitoring in personalized healthcare through digital twins. In: Proceedings of the 29th Annual International Conference on Computer Science and Software Engineering, pp. 329–335 (2019)
23. Vallée, A.: Envisioning the future of personalized medicine: role and realities of digital twins. J. Med. Internet Res. **26**, e50204 (2024)
24. Andaluz, V.H., Chicaiza, F.A., Gallardo, C., Quevedo, W.X., Varela, J., Sánchez, J.S., Arteaga, O.: Unity3D-MatLab simulator in real time for robotics applications. In: Augmented Reality, Virtual Reality, and Computer Graphics: Third International Conference, AVR 2016, Lecce, Italy, June 15–18, 2016. Proceedings, Part I 3, pp. 246–263. Springer (2016)
25. Akharas, I., Hennessey, M.P., Tornoe, E.J.: Simulation and visualization of dynamic systems in virtual reality using solidworks, MATLAB/Simulink, and unity. In: ASME International Mechanical Engineering Congress and Exposition, vol. 84546, p. V07AT07A039. American Society of Mechanical Engineers (2020)
26. Jangraw, D.C., Johri, A., Gribetz, M., Sajda, P.: NEDE: an open-source scripting suite for developing experiments in 3D virtual environments. J. Neurosci. Methods. **235**, 245–251 (2014)
27. Hu, X.: Design of virtual experiment teaching of inorganic chemistry in colleges and universities based on unity3D. Int. J. Adv. Comput. Sci. Appl. **14**(4) (2023)
28. Leskovský, R., Kučera, E., Haffner, O., Rosinová, D.: Proposal of digital twin platform based on 3d rendering and IIOT principles using virtual/augmented reality. In: 2020 Cybernetics & Informatics (K&I), pp. 1–8. IEEE (2020)

29. Rosales, J., Deshpande, S., Anand, S.: IIoT based augmented reality for factory data collection and visualization. Proc. Manufac. **53**, 618–627 (2021)
30. Kollatsch, C., Schumann, M., Klimant, P., Wittstock, V., Putz, M.: Mobile augmented reality based monitoring of assembly lines. Proc. Cirp. **23**, 246–251 (2014)
31. Abd Elaziz, M., Al-qaness, M.A.A., Dahou, A., Al-Betar, M.A., Mohamed, M.M., El-Shinawi, M., Ali, A., Ewees, A.A.: Digital twins in healthcare: applications, technologies, simulations, and future trends. Wiley Interdisc. Rev. Data Mining Knowl. Discov., e1559
32. Katsoulakis, E., Wang, Q., Huanmei, W., Shahriyari, L., Fletcher, R., Liu, J., Achenie, L., et al.: Digital twins for health: a scoping review. NPJ Digit. Med. **7**(1), 77 (2024)
33. Mihai, S., Yaqoob, M., Hung, D.V., Davis, W., Towakel, P., Raza, M., Karamanoglu, M., et al.: Digital twins: a survey on enabling technologies, challenges, trends and future prospects. IEEE Commun. Surv. Tutor. **24**(4), 2255–2291 (2022)
34. Van Der Burg, S., Kloppenburg, S., Kok, E.J., Van Der Voort, M.: Digital twins in agri-food: societal and ethical themes and questions for further research. NJAS Impact Agric. Life Sci. **93**(1), 98–125 (2021)

MIX
Papier aus verantwortungsvollen Quellen
Paper from responsible sources
FSC® C105338

If you have any concerns about our products,
you can contact us on
ProductSafety@springernature.com

In case Publisher is established outside the EU,
the EU authorized representative is:
Springer Nature Customer Service Center GmbH
Europaplatz 3, 69115 Heidelberg, Germany

Printed by Libri Plureos GmbH
in Hamburg, Germany